TECHNOLOGY: A REIGN OF BENEVOLENCE AND DESTRUCTION

Edited by
Fred A. Olsen
Western Washington State College

MSS Information Corporation
655 Madison Avenue, New York, N.Y. 10021

This is a custom-made book of readings prepared for the courses taught by the editor, as well as for related courses and for college and university libraries. For information about our program, please write to:

MSS INFORMATION CORPORATION
655 Madison Avenue
New York, New York 10021

MSS wishes to express its appreciation to the authors of the articles in this collection for their cooperation in making their work available in this format.

Library of Congress Cataloging in Publication Data

Olsen, Fred A   comp.
    Technology: a reign of benevolence and destruction.

    1. Technology — Social aspects — Addresses, essays, lectures. I. Title
T14.5.047   301.24'3'08   73-16385
ISBN 0-8422-5130-8
ISBN 0-8422-0356-7 (pbk.)

T
14.5
.047

CONTENTS

Introduction   5

HISTORICAL PERSPECTIVE   7

Agricola *De Re Metallica*--1555
    HERBERT CLARK HOOVER and LOU HENRY HOOVER . . .   8
The Ideal Training of an American Boy--1894
    THOMAS DAVIDSON   16
The Bubonic Plague--1897
    VICTOR C. VAUGHAN   21
Forecasting the Progress of Invention--1897
    WILLIAM BAXTER, JR.   23
The Utilization of Waste
    PETER T. AUSTEN   27
The Metric System and International Commerce--1901
    JAMES HOWARD GORE   38
The Condition of Wage-Earning Women
    CLARE DE GRAFFENRIED   42
A Forgotten Industrial Experiment
    SARA A. UNDERWOOD   51

CONTEMPORARY PERSPECTIVE:
THE DISORDERED TECHNOLOGY   57

Is Technology the Cause of the World's Problems?
    ROY V. HUGHSON   58
"As Long As It Doesn't Kill Anybody . . ."
    CHARLES F. WURSTER   64
In Defense of DDT and Other Pesticides
    NORMAN E. BORLAUG   75
Heresy in the Hinterland--1966
    D. NESLO   85
Lead, the Inexcusable Pollutant
    PAUL P. CRAIG   88

America's Answer to the Population Crisis
    FRED A. OLSEN 93

What Price Ecology?
    KENNETH S. TOLLETT............................. 100

The Tragedy of the Commons
    GARRETT HARDIN 103

Man's Participatory Evolution
    RENE DUBOS 115

**CONTEMPORARY PERSPECTIVE:
THE ORDERED TECHNOLOGY**   **121**

Commitment to Humanity
    R. BUCKMINSTER FULLER.......................... 122

Women and Work (III): The Effects of Technological Change
    MADELEINE GUILBERT 136

The Culture of Machine Living
    MAX LERNER 143

Women and Technology in Developing Countries
    BARBARA E. WARD................................ 148

Schools Are Going Metric
    FRED J. HELGREN 157

Technology, Technique, and the Jesus Movement
    W. FRED GRAHAM 160

Technology and Society: A Challenge to Private Enterprise
    IAN K. MacGREGOR 168

How to Love the Land and Live with your Love
    RUTH C. ADAMS................................... 178

FUTURE PERSPECTIVE   **185**

Technology in the United States: The Options Before Us
    J. HERBERT HOLLOMON........................... 186

On Making the Future Safe for Mankind
    E. J. MISHAN 207

Education for the Future
    HERBERT J. MULLER 236

Bibliography 249

INTRODUCTION

In an article drawn from his book, *The Pentagon of Power*, Lewis Mumford stated:

> We face now the great paradox of automation, put once and for all in Goethe's fable of the Sorcerer's Apprentice. Our civilization has found a magic formula for setting industrial and academic brooms and pails of water to work by themselves, in ever increasing quantities at ever increasing speed. But we have lost the Master Magician's spell for altering the tempo of this process or halting it when it ceases to serve human functions and purposes, though this formula (foresight and feedback) is written plainly on every organic process.
>
> As a result we are already, like the apprentice, beginning to drown in the flood. The moral should be plain: unless one has the power to stop an automatic process — and if necessary reverse it — one had better not start it. . . .

Over 40 years earlier and in a similar vein, Raymond Fosdick wrote:

> Humanity stands to-day in a position of unique peril. An unanswered question is written across the future: Is man to be the master of the civilization he has created, or is he to be its victim? Can he control the forces which he has himself let loose? Will this intricate machinery which he has built up and this vast body of knowledge which he has appropriated be the servant of the race, or will it be a Frankenstein monster that will slay its own maker. In brief, has man the capacity to keep up with his own machines?

Though over 40 years of history separate the two quotations, they serve to illustrate the endless concern that Man has had, does have, and will continue to have over the association between Man and his technology.

Nor is the problem restricted to this century. There are numerous instances from throughout recorded history wherein similar expressions of societal/technological concern were voiced. Two well-known examples will suffice.

There were the textile machine wreckers of early 19th century England — the Luddites.

Renaissance man George Agricola described in 1555 the adverse ecological impact mining processes were having upon the Italian countryside.

It is toward an examination of those concerns of Man and his technology that this book of readings has been compiled. The selections are those of economists, historians, philosophers, political scientists, sociologists, and technologists. Thus, students in the preceding disciplines may "come to grips" with a variety of contemporary, technological issues. An understanding of the technological issues will develop *pari passu* with the inductive analysis of the readings bearing upon any given issue.

The book has been divided into three general parts within each of which a variety of cogent opinions and facts have been expressed. The general parts are:

1. Historical perspective
2. Contemporary perspective
   a. The disordered technology
   b. The ordered technology
3. Future perspective

A word to the reader. Students in engineering, science, and technology might well examine philosophical and sociological concepts and their relationship to individual specializations. Students in sociology may find it more purposeful to examine the correlation between science (discovery) and technology (application) and the responsibilities of each. In other words, in each case, critically pursue and conceptualize those points of view of writers outside your speciality. While you may not become a Buckminster Fuller "comprehensivist" overnight, at least you will be better able to understand and generalize upon the interrelatedness of diverse issues. Ultimately, it is through acquisition of this breadth of understanding that one may most humanely formulate personal philosophy, life style, and subsequent purposeful action.

HISTORICAL PERSPECTIVE

AGRICOLA DE RE METALLICA--1555

Herbert Clark Hoover and
Lou Henry Hoover

Georgius Agricola was born at Glauchau, in Saxony, on March
24th, 1494, and therefore entered the world when it was still
upon the threshold of the Renaissance; Gutenberg's first book
had been printed but forty years before; the Humanists had but
begun that stimulating criticism which awoke the Reformation;
Erasmus, of Rotterdam, who was subsequently to become Agricola's
friend and patron, was just completing his student days. The
Reformation itself was yet to come, but it was not long delayed,
for Luther was born the year before Agricola, and through him
Agricola's homeland became the cradle of the great movement;
nor did Agricola escape being drawn into the conflict. Italy,
already awake with the new classical revival, was still a busy
workshop of antiquarian research, translation, study, and publication, and through her the Greek and Latin Classics were
only now available for wide distribution. Students from the
rest of Europe, among them at a later time Agricola himself,
flocked to the Italian Universities, and on their return infected their native cities with the newly-awakened learning.
At Agricola's birth Columbus had just returned from his great
discovery, and it was only three years later that Vasco Da Gamma
rounded Cape Good Hope. Thus these two foremost explorers had
only initiated that greatest period of geographical expansion
in the world's history. A few dates will recall how far this
exploration extended during Agricola's lifetime. Balboa first
saw the Pacific in 1513; Cortes entered the City of Mexico in
1520, Magellan entered the Pacific in the same year; Pizarro
penetrated into Peru in 1528; De Soto landed in Florida in 1539,
and Potosi was discovered in 1546. Omitting the sporadic settlement on the St. Lawrence by Cartier in 1541, the settlement
of North America did not begin for a quarter of a century after
Agricola's death. Thus the revival of learning, with its train
of Humanism, the Reformation, its stimulation of exploration
and the re-awakening of the arts and sciences, was still in its
infancy with Agricola.

. . . .

As to Agricola's contribution to the sciences of mining and
metalurgy, De Re Metallica speaks for itself. While he describes,
for the first time, scores of methods and processes, no one
would contend that they were discoveries or inventions of his
own. They represent the accumulation of generations of exper-
ence and knowledge; but by him they were, for the first time,
to receive detailed and intelligent exposition. Until
Schlüter's work nearly two centuries later, it was not excelled.
There is no measure by which we may gauge the value of such a

DE RE METALLICA, Dover Publications, Inc., 1950.

work to the men who followed in this profession during centuries, nor the benefits enjoyed by humanity through them.

That Agricola occupied a very considerable place in the great awakening of learning will be disputed by none except by those who place the development of science in rank far below religion, politics, literature, and art. Of wider importance than the details of his achievements in the mere confines of the particular science to which he applied himself, is the fact that he was the first to found any of the natural sciences upon research and observation, as opposed to previous fruitless speculation. The wider interest of the members of the medical profession in the development of their science than that of geologists in theirs, has led to the aggrandizement of Paracelsus, a contempory of Agricola, as the first in deductive science. Yet no comparative study of the unparalleled egotistical ravings of this half-genius, half-alchemist, with the modest sober logic and real research and observation of Agricola, can leave a moment's doubt as to the incomparably greater position which should be attributed to the latter as the pioneer in building the foundation of science by deduction from observed phenomena. Science is the base upon which is reared the civilization of today, and while we give daily credit to all those who toil in the superstructure, let none forget those men who laid its first foundation stones. One of the greatest of these was Georgius Agricola.

. . . .

But let us now approach the subject we have undertaken. Since there has always been the greatest disagreement amongst men concerning metals and mining, some praising, others utterly condemning them, therefore I have decided that before imparting my instruction, I should carefully weigh the facts with a view to discovering the truth in this matter.

So I may begin with the question of utility, which is a two-fold one, for either it may be asked whether the art of mining is really profitable or not to those who are engaged in it, or whether it is useful or not to the rest of mankind. Those who think mining of no advantage to the men who follow the occupation assert, first, that scarcely one in a hundred who dig metals or other such things derive profit therefrom; and again, that miners, because they entrust their certain and well-established wealth to dubious and slippery fortune, generally deceive themselves, and as a result, impoverished by expenses and losses, in the end spend the most bitter and most miserable of lives. But persons who hold these views do not perceive how much a learned and experienced miner differs from one ignorant and unskilled in the art. The latter digs out the ore without any careful discrimination, while the former first

assays and proves it, and when he finds the veins either too
narrow and hard, or too wide and soft, he infers therefrom that
these cannot be mined profitably, and so works only the approved
ones. What wonder then if we find the incompetent miner suffers
loss, while the competent one is rewarded by an abundant return
from his mining? The same thing applies to husbandmen. For
those who cultivate land which is alike arid, heavy, and barren,
and in which they sow seeds, do not make so great a harvest as
those who cultivate a fertile and mellow soil and sow their
grain in that. And since by far the greater number of miners
are unskilled rather than skilled in the art, it follows that
mining is a profitable occupation to very few men, and a source
of loss to many more. Therefore the mass of miners who are
quite unskilled and ignorant in the knowledge of veins not infrequently lose both time and trouble. Such men are accustomed
for the most part to take to mining, either when through being
weighted with the fetters of large and heavy debts, they have
abandoned a business, or desiring to change their occupation,
have left the reaping-hook and plough; and so if at any time
such a man discovers rich veins or other abounding mining produce, this occurs more by good luck than through any knowledge
on his part. We learn from history that mining has brought
wealth to many, for from old writings it is well known that
prosperous Republics, not a few kings, and many private persons,
have made fortunes through mines and their produce....
. . . .

   The critics say further that mining is a perilous occupation
to pursue, because the miners are sometimes killed by the pestilential air which they breathe; sometimes their lungs rot away;
sometimes the men perish by being crushed in masses of rock;
sometimes, falling from the ladders into the shafts, they break
their arms, legs, or necks; and it is added there is no compen-
sation which should be thought great enough to equalize the
extreme dangers to safety and life. These occurrences, I confess, are of exceeding gravity, and moreover, fraught with
terror and peril, so that I should consider that the metals
should not be dug up at all, if such things were to happen very
frequently to the miners, or if they could not safely guard
against such risks by any means. Who would not prefer to live
rather than to possess all things, even the metals? For he who
thus perishes possesses nothing, but relinquishes all to his
heirs. But since things like this rarely happen, and only in
so far as workmen are careless, they do not deter miners from
carrying on their trade any more than it would deter a carpen-
ter from his because one of his mates has acted incautiously
and lost his life by falling from a high building....
. . . .

But besides this, the strongest argument of the detractors
is that the fields are devastated by mining operations, for
which reason formerly Italians were warned by law that no one
should dig the earth for metals and so injure their very fertile fields, their vineyards, and their olive groves. Also
they argue that the woods and groves are cut down, for there
is need of an endless amount of wood for timbers, machines,
and the smelting of metals. And when the woods and grove
are felled, then are exterminated the beasts and birds, very
many of which furnish a pleasant and agreeable food for man.
Further, when the ores are washed, the water which has been
used poisons the brooks and streams, and either destroys the
fish or drives them away. Therefore the inhabitants of these
regions, on account of the devastation of their fields, woods,
groves, brooks and rivers, find great difficulty in procuring
the necessaries of life, and by reason of the destruction of
the timber they are forced to greater expense in erecting
buildings. Thus it is said, it is clear to all that there is
greater detriment from mining than the value of the metals
which the mining produces....
. . . .

But what need of more words? If we remove metals from the
service of man, all methods of protecting and sustaining health
and more carefully preserving the course of life are done away
with. If there were no metals, men would pass a horrible and
wretched existence in the midst of wild beasts; they would return to the acorns and fruits and berries of the forest. They
would feed upon the herbs and roots which they plucked up with
their nails. They would dig out caves in which to lie down at
night, and by day they would rove in the woods and plains at
random like beasts, and inasmuch as this condition is utterly
unworthy of humanity, with its splendid and glorious natural
endowment, will anyone be so foolish or obstinate as not to
allow that metals are necessary for food and clothing and that
they tend to preserve life?

Moreover, as the miners dig almost exclusively in mountains
otherwise unproductive, and in valleys invested in gloom, they
do either slight damage to the fields or none at all. Lastly,
where woods and glades are cut down, they may be sown with
grain after they have been cleared from the roots of shrubs and
trees. These new fields soon produce rich crops, so that they
repair the losses which the inhabitants suffer from increased
cost of timber. Moreover, with the metals which are melted
from the ore, birds without number, edible beasts and fish can
be purchased elsewhere and brought to these mountainous regions.
. . . .

11

It remains for me to speak of the ailments and accidents
of miners, and of the methods by which they can guard against
these, for we should always devote more care to maintaining our
health, that we may freely perform our bodily functions, than
to making profits. Of the illnesses, some effect the joints,
others attack the lungs, some the eyes, and finally some are
fatal to men.

Where water in shafts is abundant and very cold, it frequently injures the limbs, for cold is harmful to the sinews. To
meet this, miners should make themselves sufficiently high boots
of rawhide, which protect their legs from the cold water; the
man who does not follow this advice will suffer much ill-health,
especially when he reaches old age. On the other hand, some
mines are so dry that they are entirely devoid of water, and
this dryness causes the workmen even greater harm, for the dust
which is stirred and beaten up by digging penetrates into the
windpipe and lungs, and produces difficulty in breathing....
If the dust has corrosive qualities, it eats away the lungs,
and implants consumption in the body; hence in the mines of the
Carpathian Mountains women are found who have married seven
husbands, all of whom this terrible consumption has carried
off to a premature death. At Altenberg in Meissen there is
found in the mines black pompholyx, which eats wounds and ul-
cers to the bone; this also corrodes iron, for which reason
the keys of their sheds are made of wood. Further, there is a
certain kind of cadmia which eats away the feet of the workmen
when they have become wet, and similarly their hands, and injures their lungs and eyes. Therefore, for their digging they
should make for themselves not only boots of rawhide, but
gloves long enough to reach to the elbow, and they should
fasten loose veils over their faces; the dust will then neither
be drawn through these into their windpipes and lungs, nor will
it fly into their eyes. Not dissimilarly, among the Romans the
makers of vermilion took precautions against breathing its fatal
dust.

Stagnant air, both that which remains in a shaft and that
which remains in a tunnel, produces a difficulty in breathing;
the remedies for this evil are the ventilating machines which
I have explained above. There is another illness even more
destructive, which soon brings death to men who work in those
shafts or levels or tunnels in which the hard rock is broken
by fire. Here the air is infected with poison, since large
and small veins and seams in the rocks exhale some subtle poison from the minerals, which is driven out by the fire, and
this poison itself is raised with the smoke not unlike pompholyx, which clings to the upper part of the walls in the
works in which ore is smelted. If this poison cannot escape
from the ground, but falls down into the pools and floats on

their surface, it often causes danger, for if at any time the water is disturbed through a stone or anything else, these fumes rise again from the pools and thus overcome the men, by being drawn in with their breath; this is even much worse if the fumes of the fire have not yet all escaped. The bodies of living creatures who are infected with this poison generally swell immediately and lose all movement and feeling, and they die without pain; men even in the act of climbing from the shafts by the steps of ladders fall back into the shafts when the poison overtakes them, because their hands do not perform their office, and seem to them to be round and spherical, and likewise their feet. If by good fortune the injured ones escape these evils, for a little while they are pale and look like dead men. At such times, no one should descend into the mine or into the neighbouring mines, or if he is in them he should come out quickly. Prudent and skilled miners burn the piles of wood on Friday, towards evening, and they do not descend into the shafts nor enter the tunnels again before Monday, and in the meantime the poisonous fumes pass away.

There are also times when a reckoning has to be made with Orcus, for some metalliferous localities, though such are rare, spontaneously produce poison and exhale pestilential vapour, as is also the case with some openings in the ore, though these more often contain the noxious fumes. In the towns of the plains of Bohemia there are some caverns which, at certain seasons of the year, emit pungent vapours which put out lights and kill the miners if they linger too long in them. Pliny, too, has left a record that when wells are sunk, the sulphurous or aluminous vapours which arise kill the well-diggers, and it is a test of this danger if a burning lamp which has been let down is extinguished. In such cases a second well is dug to the right or left, as an air-shaft, which draws off these noxious vapours. On the plains they construct bellows which draw up these noxious vapours and remedy this evil; these I have described before.

Further, sometimes workmen slipping from the ladders into the shafts break their arms, legs, or necks, or fall into the sumps and are drowned; often, indeed, the negligence of the foreman is to blame, for it is his special work both to fix the ladders so firmly to the timbers that they cannot break away, and to cover so securely with planks the sumps at the bottom of the shafts, that the planks cannot be moved nor the men fall into the water; wherefore the foreman must carefully execute his own work. Moreover, he must not set the entrance of the shaft-house toward the north wind, lest in winter the ladders freeze with cold, for when this happens the men's hands become stiff and slippery with cold, and cannot perform their office of holding. The men, too, must be careful that, even if none

of these things happen, they do not fall through their own
carelessness.

Mountains, too, slide down and men are crushed in their fall
and perish. In fact, when in olden days Rammelsberg, in Goslar,
sank down, so many men were crushed in the ruins that in one
day, the records tell us, about 400 women were robbed of their
husbands. And eleven years ago, part of the mountain of Altenberg, which had been excavated, became loose and sank, and suddenly crushed six miners; it also swallowed up a hut and one
mother and her little boy. But this generally occurs in those
mountains which contain venae cummulatae. Therefore, miners
should leave numerous arches under the mountains which need
support, or provide underpinning. Falling pieces of rock also
injure their limbs, and to prevent this from happening, miners
should protect the shafts, tunnels, and drifts.

The venomous ant which exists in Sardinia is not found in
our mines. This animal is, as Solinus writes, very small and
like a spider in shape; it is called solifuga, because it shuns
(fugit) the light (solem). It is very common in silver mines;
it creeps unobserved and brings destruction upon those who im-
prudently sit on it. But, as the same writer tells us, springs
of warm and salubrious waters gush out in certain places, which
neutralise the venom inserted by the ants.

In some of our mines, however, though in very few, there are
other pernicious pests. These are demons of ferocious aspect,
about which I have spoken in my book De Animantibus Subter-
raneis. Demons of this kind are expelled and put to flight by
prayer and fasting.

Some of these evils, as well as certain other things, are
the reason why pits are occasionally abandoned. But the first
and principal cause is that they do not yield metal, or if, for
some fathoms, they do bear metal they become barren in depth.
The second cause is the quantity of water which flows in; sometimes the miners can neither divert this water into the tunnels,
since tunnels cannot be driven so far into the mountains, or
they cannot draw it out with machines because the shafts are
too deep; or if they could draw it out with machines, they do
not use them, the reason undoubtedly being that the expenditure
is greater than the profits of a moderately poor vein. The
third cause is the noxious air, which the owners sometimes cannot overcome either by skill or expenditure, for which reason
the digging is sometimes abandoned, not only of shafts, but
also of tunnels. The fourth cause is the poison produced in
particular places, if it is not in our power either completely
to remove it or to moderate its effects. This is the reason
why the caverns in the Plain known as Laurentius used not to
be worked, though they were not deficient in silver. The fifth

cause are the fierce and murderous demons, for if they cannot be expelled, no one escapes from them. The sixth cause is that the underpinnings become loosened and collapse, and a fall of the mountain usually follows; the underpinnings are then only restored when the vein is very rich in metal. The seventh cause is military operations. Shafts and tunnels should not be re-opened unless we are quite certain of the reasons why the miners have deserted them, because we ought not to believe that our ancestors were so indolent and spiritless as to desert mines which could have been carried on with profit. Indeed, in our own days, not a few miners, persuaded by old women's tales, have re-opened deserted shafts and lost their time and trouble. Therefore, to prevent future generations from being led to act in such a way, it is advisable to set down in writing the reason why the digging of each shaft or tunnel has been abandoned, just as it is agreed was once done at Freiberg, when the shafts were deserted on account of the great inrush of water.

THE IDEAL TRAINING OF AN AMERICAN BOY--1894

Thomas Davidson

In the American education of to-day there are two things
which force themselves upon our attention: (1) that it is in a
chaotic condition; (2) that this condition is, in the main, due
to our having no definite notion of what education is aiming at.
. . . .

When, therefore, we speak of a system of education for Americans, we do not mean merely a system suited to their needs as
members of that state whose visible centre is Washington, D.C.,
but a system suited to eternal spirits living under social and
political conditions more favorable than ever existed before
for their self-unfolding and self-realization. It is just the
existence of these conditions, and this alone, that confers
upon America all the worth it possesses, and gives it a valid
claim to our highest moral enthusiasm. It is simply and solely
because, for the first time in the world's history, it offers
the conditions under which men, by realizing the divinity latent
in them, may become absolutely free, each a law unto himself,
that it has its supreme claim upon us as moral beings. It is
this, and nothing less, that is the American ideal: it is this
that must, sooner or later, shape our entire educational system.

But it will be said, Such an ideal is not merely American:
it is universal and human. Of course it is: this is just what
the American ideal ought to be. It is pure folly to try to cultivate an American provincialism, something which, like Galli-
cism and Anglicism, shall be less than universal humanity. If
America is to perform the part assigned to her in history, she
must stand for ideal humanity and compel all partial ideals to
converge and lose themselves in hers. Her citizens must be
morally autonomous, regarding all institutions as servants, not
as masters; as expressions of their own freedom; as instruments
for the realization of greater freedom. In the world hitherto,
in spite of the fundamental teachings of Christianity, man has
been, for the most part, a thrall, owning obedience to a law
conceived to be external to him, and other than the expression
of his own true nature. In a word, he has been heteronomous.
In American life it must no longer be so. The true American
must worship the inner God, recognized as his own deepest and
eternal self; not an outer God, regarded as something different
from himself. And one need not be much of a prophet to see that
this is the goal to which, with all our blindness and all our
faults, we are steadily tending. Ideal Americanism means absolute autonomy.

We may now, therefore, put the question in this form: By
what education may a boy be prepared for complete moral autonomy,

THE FORUM, July, 1894, vol. 17, 571-581.

and for playing a worthy part in an order of things intended
to guide every human being to the same? Put in this form, the
question does not seem difficult to answer. The essential conditions of moral autonomy are easily assignable. They are:
(1) well-arranged, practical knowledge of men and things; (2)
healthy, well-distributed affections; (3) a ready will, loyal
to such knowledge and such affections. To realize these, then,
must be the aim of American education.

It will be readily seen that from this education two things
are excluded, namely, erudition and professional training.
Valuable and necessary as these are, they form no part of the
education of the American as American, or of man as man. The
subject now under consideration is the education of the citizen
and of the man as such.

But we have still before us the question, How shall this result be reduced to practice? And here let us simplify matters
as much as possible. Let us suppose that we have to deal with
parents whose children are thoroughly healthy, and who are not
only able, but also willing, to give them the best of educa-
tions. Let us suppose that such parents come to us and say:
"We are convinced that your ideal of education is the true one,
but we do not see exactly by what method and means it is to be
realized." What shall we say to them? What advice shall be
given them about governesses, tutors, home-education, schools,
and the like? To answer these questions fully would carry us
far beyond the limits of the present articles. All that can be
done here is to indicate certain guiding principles.

As our aim is moral autonomy, and this rests upon intelligence, well-distributed affection or interest, and ready will,
our efforts must be directed to the harmonious cultivation of
these. And all experience, I believe, shows that, in the early
stages of education, this can be far better carried out in the
family, under the eyes of parents, than in the school, or even
in the public kindergarten. A child's character is practically
formed, for good or for evil, during the first seven years of
its life, and it is then that the utmost thoughtfulness and
watchfulness on the part of parents are demanded. During this
time, education should, as far as possible, be unconscious, and
therefore should be carried on by those methods and means which
may be applied unconsciously. A child ought never to learn
consciously anything that it can learn unconsciously; never be
instructed in anything that it can acquire by imitation or
habituation. It is the failure to observe this principle that
is the crowning defect of the Froebelian kindergarten system
as it is practised in America. It may be set down, as a general truth, that all knowledge or habit consciously acquired is
prosaic, insecure, and dead, compared with that which comes

through unconscious imitation. It is the latter that is the
storehouse of poetry.

. . . .

When a child has reached the age of seven, the parents should
seek to combine with other parents holding views of education
similar to their own, in order to establish a small private
school, to be directed by a competent teacher, standing in intimate and confidential relations with them. This school the
child ought to attend with the utmost regularity for the next
four or five years, for a number of hours (varying with the
seasons) daily. The aims of the education imparted during these
years will be four: (1) to bring the child into noble and kindly relations to other children, enabling it to practise gener-
csity and self-control; (2) to strengthen its body and its social instincts by healthy, not over-boisterous games; (3) to
develop its memory; (4) to put it in possession of the means of
future education, reading, writing, manual facility (including
drawing), and the elements of music. Whatever is imparted beyond these should be taught in connection with the lessons in
reading, memory-exercise, and manual training.

. . . .

If children have been properly trained and instructed in the
family and the small private school, they ought to be ready, at
the age of eleven or twelve, to attend a large school, private
or public, and to do so without any detriment to their feelings,
manners, and morals. That such schools have many advantages is
certain: that they have many drawbacks is equally so. In the
case of boys more especially, the choice lies between them and
private tutors, and to which of them the preference shall be
given must depend largely upon what relation parents desire
their sons to hold to life. If, like most American parents,
they wish their boys to be good solid citizens of the current
stamp, with ordinary social, economic, and political interests, and with what they would call a wholesome dread of any
departure from the code of respectability accepted by their
class, they will send them to large schools, where, in contact
with other boys, they are pretty sure to be cured of any notions or ideals that rise above the average, or in any way de-
part from it. If, on the other hand, as is rarely the case,
parents desire to develope the individuality of their sons, to
place them beyond the influence of current opinion, and so to
aid in rendering them morally autonomous, they will place them
in charge of carefully selected private tutors, and allow them
to travel for a considerable portion of the years from twelve
to eighteen, both in their own country and abroad.

If, now, as I have tried to show, moral autonomy, resting on
large experience and wide, well-distributed sympathies, is the

ideal of American life, and therefore of American education, there can be no doubt in which direction the choice ought to fall. The tutor and travel are in all cases to be preferred to the large school. Of course the tutor must be a man of high culture and character, and bent upon developing the same in his pupils; and the plan of travel must be carefully arranged, so as not to conflict with the programme of study. Moreover, many of the best results of public-school life may be secured if two or three well-trained boys travel under the care of one tutor or of different tutors. It ought to be observed that foreign travel with American tutors is something very different from foreign residence with foreign tutors. The latter is always to be strongly deprecated, as tending to render boys not only un-patriotic, but also insensible to what constitutes the worth of American life. There is nothing which so sharply distinguishes American boys from all others as their freshness and purity of life, and these qualities are likely to suffer from contact with Europeans--especially with French and German--boys. A European-ized American is nearly always a moral eunuch.

. . . .

In saying that the years from twelve to eighteen in the life of a boy whose parents desire for him the ideal American culture should be spent with a tutor in travel, I do not mean either that the whole of this travel should be in foreign countries, or that it should not be broken by periods of home-study. The contrary, indeed, is my opinion. A boy ought never to be long withdrawn from his country or his home.

I have said that most boys who find themselves in a position to do so will go to college about the age of eighteen. Nor will they be wrong in doing so: for, notwithstanding all the draw-backs of our colleges; their want of definite unitary aim and ideal; their half-mediaeval, half-professional curricula; the dry, uninspiring formalism and useless erudition of much of the teaching; the easy philistinism or dreary pessimism of many of the older teachers; the pert, callow Germanism of many of the younger ones; the boyishness of many of the students; their smug foppishness and stupid devotion to half-brutal games and half-silly girls,--notwithstanding all this and much more, there are elements in college life which the youth who aims at free man-hood cannot afford to overlook. It is at college that the young man who by discipline, study, and travel has attained self-control, an earnest view of life, and a large, generous outlook, can best put these qualities to a practical test, by mingling, on free and easy terms, with men of his own age, entering into intimate relations with them, and comparing their aims with his own. He is now old enough to have definite views and purposes, but not too old to learn how they must be modified in order to be serviceable in the actual world. And nowhere can he learn this sooner or better than at college.

. . . .

Since, as we have seen, the aim of education is the free,
self-directing man, the man ready and able to act intelligently,
nobly, and strongly in all the affairs of life; the man who,
while taking full account of the world in which he has to act,
does not find the principle or sanction of his action there, but
in his own righteous will,--it is obvious that the matter of education will include mainly those sciences which relate to man,
and his relations to the world of nature and the world of spirit,
--the humanities, as the good old expression was. The mathematical, physical, and philological sciences, as not answering
directly to this description, will occupy a subordinate position.
After much careful observation and inquiry I am convinced that
these sciences, important as they otherwise are, have comparatively little educational value. Mathematics are, indeed, good
"mental gymnastics," and it is certainly important to understand
the methods and main results of the physical and philological
sciences; but more cannot be said.

. . . .

I am aware that the ideal which I have set up in this article
is high and unworldly; but I am sure that it is the true American ideal, and I know that it has been already realized by not
a few of our young men. For it should never be forgotten that,
in spite of all that is said, and justly said, abroad, about the
irreverence and flippancy of "Yankee" children, the noblest
types of young manhood, as well as of young womanhood, that the
world knows, are to be found in America.

THE BUBONIC PLAGUE--1897

Victor C. Vaughan

Those twin monsters of human misery, Famine and Disease, are
now holding high carnival in India. Death follows in their
wake and gathers in a rich harvest. Appeals to the charitable
of the world are being made, and the civilized nations of Europe and America are looking apprehensively toward the East.
The great plague, which has confined its ravages for the most
part to certain limited districts of Asia for the past two
hundred years, seems to have grown strong enough to threaten
to take a journey abroad. The black death has unfurled its
banner in the face of modern civilization....

. . . .

Cantline makes the following statement concerning the susceptibility of rats to the disease: "On all hands rats are
reported to behave peculiarly and with a wonderful constancy.
Before, or it may be during an epidemic of plague, or before
the individuals in any particular house in an infected locality
are stricken, the rats leave their haunts and seek the interior
of the house. They become careless of the presence of man, and
run about in a dazed way with a peculiar limping jerk or spasm
of their hind legs. They are frequently found on the bedroom
floor or on the tables, but not infrequently their death is
known by the putrefactive odor of their decomposition arising
from beneath the flooring. So pertinent has this rat affection become, that during the epidemic in Macao in 1895 the
Chinese and Portuguese left their houses when the diseased rats
invaded their premises. The cause of the rats' behavior is undoubtedly disease, and the symptoms tally wonderfully with
plague symptoms of man. Dr. Rennie examined them carefully in
Canton, and noted the following post-mortem appearances: (1)
the stomach was distended and filled with particles of food,
sand, and indigestible substances, and the mucous membrane was
red and inflamed toward the pyloric end; (2) the liver was much
enlarged and congested, and contained ova of taenia and distoma;
(3) there was congestion at the base of the lung present in
about forty per cent; and (4) glandular enlargement was present
in thirty per cent of those examined. There is no doubt now
that the disease in the rat and man is identical. The bacillus
of plague has been met with in every case of rat disease of this
description when it has been searched for. The infection of the
rat is raised from being a mere popular belief into one of
scientific precision, and we must now accept the rat, at any
rate, as one animal liable to the plague. Whether the rat is
affected previously to, coincidently with, or subsequently to
man being attacked is open to question....

. . . .

POPULAR SCIENCE MONTHLY, May, 1897, vol. 51, 62-69.

There is no known racial immunity to this disease. It is alike fatal to Mongolians, Africans, and Europeans. It has prevailed in the marshes along the Euphrates and on the Himalayas; in densely populated cities and in sparsely settled rural districts; on the sands of Egypt and amid the snows of Norway.

Climate and season have been studied in order to establish between them and the plague a causal relationship. Epidemics have followed prolonged droughts, and have prevailed during rainy seasons. The wind may blow where it listeth, but the bacillus heedeth it not. The epidemic at Hong Kong in 1894 appeared after a prolonged season of dry weather. Rain was anxiously looked for--probably prayed for. It was said, "All will be well when the rain comes." At last the rain did come, and with it the disease seemed to be refreshed and the number of deaths was multiplied. The attempt to find in meteorological conditions a cause for our ills is a relic of the superstition of ages when it was believed that disease was sent from heaven to afflict man for his sins, and was due to the anger of the gods.

Overcrowding is undoubtedly a factor in the distribution of this disease, as it is of all other infectious diseases, simply because it renders transmission of the germ from one to the other more speedy and certain; but that the disease can be due to overcrowding is, in the present state of our knowledge, an absurdity.

Poverty and famine are factors in the propagation of the disease. Want of proper food renders the individual more susceptible. This has been demonstrated in case of more than one infectious disease by experiments upon the lower animals. Privation has always been associated with the most notable outbreaks of the plagues. As stated in the beginning of this paper, famine and disease are twin brothers, inseparable. Where one of them dwells there the other may be found. This is undoubtedly the reason why this disease has always found a home in the Levantine. Cantline says: "In the densely packed cities of Asia the poor exist forever on the fringe of destitution, and the least rise in the price of food brings scarcity, so that the term, 'the poor man's plague,' holds good for all time."
. . . .

FORECASTING THE PROGRESS OF INVENTION--1897

William Baxter, Jr.

The great progress made during the last fifty years in the
domain of science and invention has aroused a very general desire among intelligent people to know what the future has in
store, and in many cases the desire has become so strong as to
develop prophetic tendencies. Whenever a banquet is given in
commemoration of some scientific event, or upon the anniversary
of some ancient and honorable society, the orator of the evening is sure to dwell at considerable length upon the great discoveries that are still to come. By contrasting the extraordinary advances made during the last century with the comparatively
limited progress of all previous time, and by showing that the
rate of advancement has been continually increasing during the
latter period, he arrives at the conclusion that in the years
to come development will increase in a compound ratio, and the
discoveries will become so numerous and so great as to dwarf
into insignificance all that has been accomplished up to the
present time.

Writers who dwell upon these glorious achievements of mankind
in modern times follow the same vein, and make equally extravagant predictions as to the future. If these writers and orators
would stop when they reach this point in their meditations they
would be wise, since it is a self-evident fact that progress in
science and invention has been increasing very rapidly during
the last fifty or sixty years, and certainly there is no reason
to suppose that we have reached the end, and that henceforth
development will be very slow; but at this point the spirit of
prophecy seizes them, and they proceed to describe the wonders
yet unseen. It is here that they almost invariably fail. They
would not be satisfied if they assumed that future progress
would be along the lines of possible development--that would be
too commonplace and altogether out of keeping with the ideal of
the greatness of the future achievements of mankind. They must
necessarily assume that what is brought forth hereafter will be
so far in advance of what we now know of as to be revolutionary
in its character, and so much so, in fact, as to consign to the
scrap heap the most perfect devices of the present time. Some
of the means by which these results are to be attained are not
capable of accomplishing such wonders; others, while of great
theoretical possibilities, are surrounded by certain practical
difficulties so well understood at the present time that we can
almost with certainty say that they will never realize the
dreams that are based upon them. The remainder are problems
that can be solved today, and would be if it were not for the
fact that it is by no means certain that their solution would
be of any practical value. The improbability of ever realizing

POPULAR SCIENCE MONTHLY, May, 1897, vol. 51, 307-314.

a substantial gain by the solution of many of the problems upon which prophecies as to the wonders of the future are based is fully appreciated by many of those who have given the subject careful consideration; but those who dream of the revolutionary character of future invention never take note of such things.

Nearly all those who succumb to the fascination of meditating upon the changes that may be wrought be inventive genius in days to come follow the same line of thought. The problems upon the solution of which their fancy paints its pictures are always the same, although some contemplate the whole category, while others only dwell upon a portion thereof. These problems are aërial navigation, the development of electric energy direct from coal or some other equally cheap substance, and the utilization of the various forces of Nature, such as solar heat, tide and wave motion, and wind currents. Of these, aërial navigation is supposed to be by far the most important, obtaining electricity direct from coal and the others following along in the order in which they are given above.

As to the utilization of solar heat, tides, wave motion, and wind currents, it can be truthfully said that they could be utilized at the present time if it were considered profitable to do so. The energy of wind currents, as every one knows, is made available on a very extensive scale, but always in small units, and this fact alone shows that it can not compete with the steam engine, which, according to the prophets, it is sure to supersede. The energy of tides and wave motion is also utilized to some extent, and solar engines have been made from time to time.

It can not be said that these unlimited sources of energy are not brought into the service of man because of our inability to devise apparatus with which to harness them successfully, for, as a matter fact, a great deal of ingenuity has been displayed in this direction, and the cost of the mechanism, with reference to the power recovered, has probably been reduced to nearly as low a point as is possible. In the matter of simplicity and durability equally good results have been obtained.

. . . .

As these natural forms of energy can be obtained without cost, and the fuel used by a steam engine has to be purchased, it is apparently reasonable to assume that they would constitute a more economical form of power, but wherever a constant supply is desired it is very doubtful whether the economy of the steam engine can be superseded by any one of them. It is true that there is no expenditure for fuel, but the interest on the extra cost of the plant and the maintenance thereof,

as well as the additional space required, may more than offset
this gain; and the fact that so little is done in the way of
utilizing them would seem to show that up to the present time
their value has failed to make any great impression upon engineers who have looked into the subject. It does not follow
from this that they will never come into use on a more extensive scale than at present, but it does follow that the dreams
of those who believe that they will eventually supersede all
forms of prime movers that consume fuel will never be realized.
Through the increased value of fuel or the reduced cost of construction of the apparatus, or both, they may become competitors to a greater or less extent, but more than this can not
be expected.

Considering, now, the effects of the solution of the problem
of obtaining electricity direct from coal, it can be said that
it is far more likely to revolutionize the affairs of the world
than the utilization of the natural forms of energy; but it
must also be said that we are not justified, in view of what is
now known in relation to the subject, in assuming that it will
ever realize the predictions of the oversanguine prophets. If
we could solve the problem according to our ideal, all that is
expected of it would be accomplished; but such a solution is
highly improbable, if not actually impossible. Our ideal battery would be as simple as a boiler, and be provided with a
place where coal could be fed in and another through which the
residue could be removed. In a boiler, the pressure of the
steam, as well as the quantity generated, can be increased by
simply increasing the size of the fire box, but this simplicity
could not be obtained even in our ideal battery, because the
electromotive force would remain the same no matter how much
the size of the cell might be increased. To obtain an electromotive force high enough for practical purposes it would be
necessary to use a large number of cells, and, to feed these
without too much trouble, it would be necessary to devise an
automatic feeder capable of operating with a degree of perfection hardly obtainable without the aid of human intelligence.
. . . .

It must not be assumed from what has been said in the fore-
going that the writer regards the solution of the problems here
considered as of no special value, for his views are just the
opposite of this. The object aimed at has been to show that
the wonderful things that it is expected will be accomplished
by the solution of these problems will never be realized with
regard to some because they are not possible, and are not
likely to be realized by the others on account of inherent
defects that the solutions may bring to light. The coal-
battery problem will, no doubt, be worked out, in some form or
other, but who can tell whether the objectionable features of

it will or will not offset all its advantages? The hot-air
engine is a far more perfect converter of energy, in theory,
than the steam engine, but its defects when reduced to a practical form are such that it is of no value except for small
power, and this may also turn out to be the case with the coal
battery. The utilization of the energy of tides, solar heat,
etc., is as possible to-day as at any future time; the fact
that they are not utilized is proof that they are not considered as desirable as other forms of energy. In the future the
cost of the apparatus for harnessing them may be so reduced as
to render them available to a much greater extent than at the
present time, but that they will ever revolutionize the industrial affairs of the world and drive the steam engine out of
use is hardly a remote possibility.

THE UTILIZATION OF WASTE.

PETER T. AUSTEN.

LIEBIG relates that when he was a young man, a manufacturer of Prussian blue, who was showing him through the works, drew his attention to the great noise made by certain comminuting and mixing machines. These machines consisted of large iron mortars in which iron pestles were actuated by machinery. The pestles pounded the materials to a fine condition and mixed them. On Liebig's suggesting that some means ought to be devised to prevent the terrible din made by the machines, the manufacturer told him that it was a singular fact that the more noise the pestles made the better was the blue produced. It happens that in making Prussian blue iron is a necessary constituent of the mixture, a fact which did not seem to be thoroughly appreciated by this manufacturer. He was, therefore, much surprised when Liebig told him that the iron which was necessary to produce the color was being rubbed off his machinery — a most extravagant way to supply it. He understood for the first time, however, why it was that the greater the noise from the friction of the pestles in the mortars, the better was the blue produced. He thus learned in an expensive way that it was better manufacturing to put iron into the mixture than to grind it off high-priced machinery.

In an experience with manufacturers extending over twenty-five years I have seen a good deal of this kind of blue manufacturing. In fact, I have seen manufacturers carrying large stocks of various kinds of blue; but in the majority of instances a dose of chemistry properly applied has removed their difficulties and turned losses into profits.

It is evident that if a certain raw product yields 1,000 pounds of a finished product and 100 pounds of a substance that is not salable and has no use, this valueless product, or waste, has still to be handled, dumped, or conveyed away, as the case may be, and that all this means expense. It may accumulate in unsightly piles, may pollute the air, may choke up or contaminate streams, or may occupy valuable ground. The cost of handling it is charged to the main product. If this 100 pounds of valueless waste can be converted into something which has a commercial value, then the returns from the sale of this article reduce

FORUM, September, 1901, vol. 32, 74-84.

the price of the main product by distributing the cost of production. It is also evident that if two manufacturers are making the same article, and one can get a price for a substance which in the other man's process must be reckoned as a valueless waste, the competition between them may be fierce, but it will be brief.

As soon as chemists were permitted to study manufacturing processes they began to invent methods of recovering substances of commercial value from the immense quantities of waste products daily produced; of increasing the yield of products; of increasing the purity of products; and of decreasing the cost of manufacture. A mere mention of the work done on these lines would fill volumes, but a reference to some of the cases may prove interesting reading. I will outline particularly some of those connected with the utilization of waste; drawing attention to applications which have been suggested as well as to those which have been technically introduced, and not omitting such inferences and relationships as may appear worth being recorded.

It is customary for writers on this topic to dilate on the truly marvellous instance of the utilization of coal tar. This fact has been emphasized so often that I shall take it for granted that my readers know that this offensive tar, which was once a waste and a nuisance, now yields on distillation a series of products each of which serves as a starting-point for a long series of valuable substances, including the beautiful aniline dyes, perfumes, medicaments, antiseptics, and what not. From the former waste product has grown up an industry in which many thousands of persons are employed, and many millions of dollars invested. It is, however, a matter of regret that although this country is an immense producer of coal-tar, and exports large quantities of tar distillates, and although the processes for making coal-tar derivatives are almost all patented here, the manufacture of them has yet been but little developed in the United States. It is a question if it is good business policy to protect by a United States patent a product which is not manufactured in this country, and on the importation of which a duty must be paid.

Garbage represents a waste which is usually very much in evidence. In civilized communities it is separated from the ashes, and hence is in a condition for utilization. Instead of being a waste and an unmitigated nuisance, garbage may now be converted into several valuable products. The most successful treatment consists in heating the garbage with water and chemicals until it is converted into "soup," removing the grease by means of raising and centrifugal apparatus, evaporating the "soup" to solid form, and grinding it. All the gases given off during the evaporation of the "soup" and the treatment of the garbage are burnt, and are thus utilized as fuel. The grease is of good quality, and finds a ready market. When purified, it can be used in making soap. The dried substance is sold as "tankage;" and as it is rich in ammonium salts, nitrogen, and phosphates, it commands a good price from the manufacturers of fertilizers.

City refuse is burnt in especially constructed furnaces. The heat produced is utilized to make steam for power or heating. Electric light and power plants are run by heat derived from the burning of garbage, and certain electrolytic chemical processes could also be run, if desired, and used to purify the drinking-water, as well as for other purposes. From the ashes a cement is made. In fact, a well-managed community ought to make its garbage and refuse pay for its removal and for some other things as well. A successful process for separating fine coal from city ashes would find a profitable opportunity in large cities. The amount of fine coal which at present is lost in the ashes of New York is estimated at from 600 to 1,000 tons a day.

The utilization of cesspool matter has been successfully worked out in the city of Augsburg. The material is treated with acid, and dried to a powder, which forms a valuable fertilizer. The value of the available nitrogen contained in this material annually lost in the city of Munich alone has been estimated at $500,000. This waste might well be worked up to replace some of the millions of dollars' worth of Peruvian guano imported yearly into Germany to manure the soil. The city of Antwerp once paid $5,000 a year to get rid of its refuse. Later on it received $200,000 a year from the same refuse; scientific men having found ways to utilize it. It was calculated in 1872 that the annual loss of valuable material through the sewers of the United States equalled a sum sufficient to pay half the interest on the national debt. This matter, however, is not altogether lost, for it is to some extent assimilated by marine life.

Waste soapsuds constitute another immense amount of material which flows from the textile factories in great streams, polluting rivers and making no end of trouble. As an instance of how this can be worked up, the yarn mills at Mülhausen may be cited. The soapsuds are precipitated with lime, and the coagulum is collected, pressed into bricks, dried, and heated in gas retorts. A gas is obtained which has three times the illuminating power of coal gas. Nearly twice as much gas can be produced as is required to light the factory. Other utilizations have also been made of waste soap liquors, such as the production of lubricating oils, fat acids, and soaps. A vexatious waste was formerly that of the liquors from making soap, which contain, among other substances, glycerine. Although the extraction of the glycerine from these waste liquors was beset with technical difficulties the problem was finally solved.

The extraction of grease and fat from dead animals and offal has also been brought to a high state of perfection. The carcases are placed in large receptacles and treated with benzine. On the evaporation of the solvent the crude fatty matters are obtained, and on purification these yield excellent materials, which appear later in the form of soap, lubricants, oils, etc.

The methods of working up some of the apparently waste objects are full of interest. Few people think of what becomes of the hundreds of horses which die every day in the streets of a large city. Practically no part of them goes to waste. I quote the following disposition of a dead horse from Simmonds:[1]

Hair: used for hair-cloth, mattresses, bags for crushing oil seeds, plumes, etc.
Hide: used for tanning and covering tables, etc.
Tendons: made into glue and gelatine.
Flesh: boiled for food for cats, dogs, and poultry.
Blood: used in manufacture of prussiate of potash, and manure.
Intestines: used for covering sausages.
Grease: used in manufacture of candles and soap.
Bones: used in making knife-handles, manure, phosphorus, and superphosphates.
Hoofs: made into pincushions and snuff-boxes when polished, or used for making glue, gelatine, and prussiate of potash.
Shoes: sold for old iron.

Some of the bones are also converted into bone charcoal, which is largely used in bleaching sugar and in medicinal substances.

Bones are used for a great variety of purposes, including the manufacture of bone char for bleaching, empyreumatic oils, tallow, black pigment for painting, shoe-blacking, and filling sheet rubber for overshoes, bone dust for manure, sulphate of ammonia, cupels, vitrified bone for use in opal glass, aside from the manufacture of knife-handles, combs, fans, buttons, etc. Bones also furnish gelatine and glue, and are the starting-point for the manufacture of phosphorus.

Rats offer a great opportunity to some one who has the knowledge and business ability to raise them. They breed easily and rapidly, and will live on garbage and offal. Their skins make a leather tougher than kid, and grace the hand of many a woman who would scream at the

[1] "Waste Products and Unapplied Substances," p. 56.

thought of a rat about her. Their fur is used by hatters, and is said to be more delicate than the beaver's.

Mastiffs furnish a good leather for boots, shoes, and riding-gloves, while the skins of small dogs and cats make a good white glove leather. Dogs, when rendered, yield a fat which is used in dressing certain kinds of glove leather. Cat's fur figures under various fancy names for women's garments.

Waste human hair occurs in larger quantities than one would suppose. In 1872 human hair to the extent of 100,000 pounds was imported from China into Marseilles. Waste hair is largely used as a manure. Heavy cloth and shawls have also been made from it. It can also be used in the manufacture of prussiate of potash, and indirectly in the production of Prussian blue. Eventually more or less of the hair finds its way into fertilizers.

There are many instances of published experiments in the utilization of waste which read as if they were intended to be facetious, although the investigators are wholly in earnest. One chemist made a wine jelly out of old boots, and was enthusiastic about it — more so, probably, than those who partook of it and did not learn the fact until later. Another converts old shirts into glucose, ferments it, distils off the alcohol, colors and flavors it, and produces a fine grade of whiskey. Another scientific experimenter feeds his whole family with cakes containing a large amount of powdered sawdust, and is pleased to find that the diet is relished and digested.

Speaking of sawdust, it is a singular fact that in the United States this waste has been allowed to accumulate in various places until it has become a nuisance, especially when it has been allowed to run into rivers and choke up the channels, disfigure the shores, and in other ways make trouble. Although there are not a few methods of utilizing sawdust, but little attempt seems to have been made in this country. The amount of sawdust produced has been reduced by the displacement of the circular saw by the band saw, which makes much less dust. There is, however, plenty of sawdust produced, and as a rule it does not find any application. In view of this state of affairs, a brief mention of some of the utilizations of sawdust will be of interest.

The first use, as would naturally occur to any one, is to burn it. Several special forms of furnaces have been invented to do this efficiently. In some cases the sawdust is dried before it is burnt, while in others it is dried in the furnace in which it is burnt. It can also be mixed with coal slack, spent dye-wood, turf, peat, etc., and compressed into briquettes, which allow it to be sold for use in place of coal. As coal is often very highly priced in regions where sawdust is produced, there is an opportunity for a large profit. By the dry distillation of sawdust all the distillation products of wood are obtained; and this manufacture can be conducted as an adjunct to the working of wood in a way to insure a profit. The products obtained are gas, wood alcohol, acetic acid, tar, and oils. From the tar there have been obtained benzole, toluole, zymole, cumole, paraffine, naphthalene, and hydrocarbons which are used in the manufacture of aniline colors. Carbolic acid and creosote are also obtained. As a last product charcoal is left in the retorts.

By sieving out the coarser particles, mixing the remainder with various fillers and agglutinants, compressing, and heating, some very interesting materials have been made, in the way of artificial wood, plastic masses, etc. Such a material was brought out by Latry in Paris, and was made from prepared sawdust and blood. It formed a hard, black substance, which could be tooled and machined like wood. It took a high polish, and could be glued, lacquered, and gilded. Imitation marbles have also been made from the fine dust of certain woods and the dust of ivory and similar waste. A mixture of sawdust and phosphate of lime with a binder has been used as a material for taking casts of sculptures, and has been called "Similibois." Slabs for parquet floors have also been made from sawdust, as well as plates for bas-reliefs, art castings, panels, and decorations. Terra-cotta lumber and artificial lumber are both instances of the utilization of sawdust. Sawdust compositions have also been used for sidewalks and dinner plates.

A long list could be given of explosives and varieties of gunpowder that have been made from sawdust. In some the sawdust is used as an absorbent, as with nitroglycerine, in others as a filler, while in still others it is converted into forms of pyroxiline. By heating sawdust with caustic alkali and sulphur, a brown dye is obtained which is cheap and fast, resisting both acids and alkalies, and dyeing cotton without a mordant. By heating sawdust with caustic alkali, oxalic acid is formed. A large amount of the oxalic acid on the market is made by this process.

There are still many other uses for sawdust. It is used to sprinkle on floors, to assist in sweeping, as a filler for fertilizers, for packing, etc. Soft-wood sawdust, mixed with slaked lime, makes a mortar which has been used for decoration. Several inventors have used mortars containing sawdust for stucco and wall finish. Mixed with cement and plaster a mass is produced which has a most remarkable insulating power against heat and cold. The spaces between the floors and walls in many of the large city structures are filled with mixtures of this nature. The material known as "Scifiarine" is made of sawdust, hemp fibre, starch, glue, and filler. It is very hard and elastic, and takes a high polish. By heating pine and fir sawdust with water and a little muriatic acid under pressure, the cellulose is converted to some extent into grape sugar, and this can be fermented. In this way brandy free from woody or turpentine odor or taste, and of excellent flavor, has been obtained from sawdust.

An ingenious use for sawdust is to mix it with clay or other non-combustible material and then burn the sawdust out, leaving a porous mass. In this way are made the alcarrazas, or porous flasks, which are used by the Arabs to cool water by evaporation on the outside. Porous bricks made in this way form walls which are admirable non-conductors of heat, as they are filled with air cells, thus utilizing the well-known non-conducting property of air. On account of its porous and consequently non-conducting nature pumice-stone was very popular with the Romans, who, when the supply of it gave out, made porous bricks by mixing clay with materials which were destroyed during the burning. Construction with porous material of this kind is not only fireproof but is very light; and hence structures can be erected on foundations which will not bear a heavy load. A mixture of sawdust, cement, and sand forms a mass which is unsurpassed as an insulating filler for walls and compartments. Certain kinds of sawdust can be pulverized and used in the manufacture of paper. Attempts have been made, but not with entire success, to make out of sawdust a substitute for cork. An excellent illuminating gas can be made from sawdust. In fact, in some localities there is enough sawdust produced to make all the gas required for the whole community, both for light and heat.

A most interesting waste product to the chemist and engineer is the slag from iron blast furnaces. Immense deposits of this material exist in various parts of the world, awaiting utilizations extensive enough to afford an outlet for them. Many attempts have been made in this direction. Some of them have been successful, but they have not as yet made much impression on the immense output of slag. Slag wool is largely used for a non-conducting and fireproof filling in construction. It is made by blowing a current of steam through a stream of melted slag. The slag is broken into little drops which fly through the air like comets, and leave a long tail, or filament, behind them. These tails constitute the slag wool. On baking this slag wool a material is formed which, it is claimed, can be used in the production of porcelain. Bricks and building stones have also been made out of slag. The use of slag for road-making is a very attractive one. As slag is a kind of glass, chemists have tried for years to modify it so as to convert it into some form of glass for which there is a large demand, as bottle glass. It is believed that very recent discoveries have made this possible in a practical degree.

In certain foreign works the heat in the slag was utilized in an interesting way, which if it had been devised in this country would have been designated as pure Yankee ingenuity. The slag was cast into slabs, and, while red-hot, was placed in receptacles in the form of travelling bags, but made of insulating material. The workman carried this home with him, and placed it in a stove made to hold the slab. In this way he got his heat without cost. The escape of heat through slag is very great, and but few attempts have been made to utilize it. Water can be boiled with it, air can be heated with it, perhaps even steam boilers might be kept going with it. Slag has also been used in making tiles, building blocks, bottles, and cement.

The immense quantity of waste gases from the iron blast furnaces was once the cause of much annoyance. It represented also a loss in heat which kept the cost of iron at a figure so high as to limit its use in many ways. Means were invented to utilize this waste in heating the air used in the blast. An immense saving was thus effected. The gas, after being thus used for heating purposes, will burn when mixed with air, and in Europe this mixture is now being used to make power. It is evidently quite inconsistent for a manufacturer to throw away large quantities of combustible gas and then buy coal to do what could be done with the gas he wastes. And before burning the gas there are some constituents which might be extracted in a way to bring about a still greater saving in cost.

The utilization of skim-milk in recent years is a good instance of the practical benefits of scientific investigation. The introduction of the Delaval separator made it possible to separate the cream from milk in a far more practical and perfect way than had been possible before. But while this aided the butter industry, there was produced a large amount of skim-milk which had little value, and, except to serve as food for hogs, had but few uses. This skim-milk contains, however, casein, albumen, and milk sugar, and can also be used in the production of lactic acid. The result of experiments has been the development of a large industry in manufacturing from the products of skim-milk coatings and sizings for paper, waterproof glues for wood veneers and other purposes, paints, substitutes for hen's eggs, hard rubber, horn, etc.

The working up of old textiles arouses many thoughts in one's mind as one sees the shop windows filled with cloth at low prices. The old clothes, consisting of mixtures of cotton and wool fibres, are exposed to the effects of various chemicals in a heated condition. The cotton fibres are thus disintegrated and when the fabric is beaten the cotton is reduced to dust and can be blown out, leaving the unchanged wool fibres. This wool is then washed, bleached, and dyed, reappearing in overcoats and many forms of cheap dress goods. The product is known in the trade as "wool extract." How many times these overcoats, trousers, vests, coats, skirts, and what not wander through the "carbonizing chambers" to reappear as new fabrics none can say. The cycle is really endless; because after the cloth has become gradually lost during its travels, or becomes too degraded even for the "carbonizer," it is converted into fertilizers, or finds its way to the soil, and so in due time reappears to a greater or less extent as wool on sheep and cotton in the boll.

To what strange uses are some materials put! Some years ago there was quite a run on dress buttons made out of blood. I do not suppose that the fair women who wore them ever dreamed of their source. It is perhaps not well to ask too many questions about the origin of things; but nothing shows more ability to grasp scientific knowledge and apply it than success in converting some worthless substance into something that is useful and valuable. But to be able to utilize waste one must have not only a thorough knowledge of the properties and changes of matter, but also a wide acquaintance with the practical applications of substances in all branches of industry; for the waste of one manufacture may be of great value to an entirely different manufacture. To find that out demands the possession of a far more extensive and intimate knowledge of technology than is usually possessed.

Some of the applications of waste are very unexpected and surprising to one not versed in this kind of work. All sorts of fantastic thoughts may be evolved in this study. Artificial pearls are made from fish scales. The old paper collars, so popular in parts of the West, find themselves again in paper. The old coffee pot is chopped up. The good parts of the metal are used for making the metal corners on trunks. The rest of it may appear as copperas, the useful mordant, or in ink. My lady writes tender sentiments to her lord with ink made from an old coffee pot, on paper made from old collars or shirts. The offal, intestines, and useless odds and ends of animal life all come handy in the production of Prussian blue, which again is the base of the greens used in paint. Bones, too, are admirable examples of queer changes. Out of bones are made phosphoric acid and acid phosphates, and from these again come some of the kinds of baking powder. Again, from these bones is made phosphorus. Among the various uses of old newspapers is the manufacturing from them of paper boards from which paper boxes are made. Old rubber shoes and bicycle tires are ground up; and the rubber is extracted and started again in its cycle.

Deposits in wine casks consist of argol, or crude bitartrate of potash, which, when purified, becomes the cream of tartar which is so popular in baking, and from which tartaric acid is made. How many of those who so earnestly oppose the use of wines, brandy, and the fermentation products of the grape understand that in using cream of tartar in raising biscuit and cake, or in wearing the innumerable fabrics dyed with tartar mordant, or in drinking effervescent salts containing tartaric acid, they are materially aiding in the support and development of the wine industry by maintaining a demand for the products of argol, which is one of its side-products? To think that persons who have been so strenuous in abolishing the army canteen should themselves be actively assisting in cheapening the cost of wine, and thus popularizing its use! Truly, where ignorance is bliss, it is folly to be consistent. If the beer industry did not supply brewer's grains at a low price for cattle food, the teetotaler might have to pay more for his milk. If the wine industry did not make a large demand for corks, where could be obtained the cork waste which serves as the basis for the manufacture of linoleum?

The strangeness of these transformations of matter is not always realized by those not familiar with chemistry. Matter is continually passing through its endless cycle. An overcoat may have in it the remains of ball-dresses and prison shirts. It may have lain on luxurious beds or in the gutters. When our shoes wear out they are made into fertilizers, and produce grass and grain, and from the grass and grain are raised cows, and out of the cow's skin we make leather again. So we have the shoe back again, less that portion of it that has been consumed as milk and beef. Nothing is really lost in Nature. Give the ground filth it returns us the flower. Matter is in eternal circulation. "Give me the sewage of New York City," says Dr. Long, "and I will return you yearly the superior milk of 100,000 cows."

I might go on for many pages telling about the waste products in manufacturing, and explaining how many of them have been utilized, and how some of them have become so valuable that they have given rise to new industries, and hence have passed from waste to raw products. Many extraordinary and unexpected relationships might be pointed out between articles of commerce, and many a singular inference drawn; but the really great lesson which I wish to teach is that in the present state of science there ought to be but very little waste. A waste product, like dirt, is matter out of place, and represents a pecuniary loss; for the matter going to waste has been bought and paid for.

To make the waste valueless the value of the material in it has to be charged for in the products sold; thus holding the price up and restricting the sales. Some of the energy and money spent in the selling of goods would show better results if spent on the reduction of the cost of making the goods which would come from a proper utilization of the waste. Finding a place for this matter out of place at once gives it a value, and takes some of the load from the cost of manufacturing the chief products. It is, therefore, a matter of great interest to the manufacturer to keep a sharp eye on his waste and to lose no opportunity to find ways to utilize it; always bearing in mind that waste is matter out of place, and that in some other branch of industry that particular form of homeless matter has a place. In fact, one manufacturer often buys what another throws away.

THE METRIC SYSTEM AND INTERNATIONAL COMMERCE--1901

James Howard Gore

When but a single man walked the earth his efforts were directed solely toward meeting his own wants. He was not called
upon to work to another's orders, but in his own necessities
he found all the measures he needed. His axe-handle or walk-
ing-stick could have served as his unit of length; and as it
would be carried directly from one article to another which
was to be measured, it was of no consequence if his crude
standard should be lost.

With the advent of the second man there arose, sooner or
later, a desire to possess the fruits of each other's labors.
But as long as the two parties to the transfer could meet face
to face, there was no confusion even when each one used his own
measure. It was the exchange of one pile for another pile, of
one group of articles for another group. As the circle of exchange grew larger, and money or a token was offered in return
for the articles desired, it became necessary for the buyer and
seller to have in mind the same sort of money and the identical
unit of measure. Thus, out of the complex relations between
man and man has arisen the need for weight and measure; and
with the forging of closer relations between men and men, or
between nations and nations, comes the necessity for the uni-
fication of all weights and all measures--the indispensable
and universal instruments of commerce.

Although all peoples appreciate the confusion that lurks in
a misleading synonomy, we find nations that are willing to continue "the application of the same generic term to different
specific things, and the misapplication of one specific term
to another specific thing." Or, to become more definite, the
English system gives us an avoirdupois pound that is heavier
than the troy pound, while the ounce avoirdupois is lighter
than the ounce troy. The ounce, the drachm, and the grain are
specific names indefinitely applied as indefinite parts of an
indefinite whole; a dozen may be 12, 14, or even 16; 28 and 25
are quarters of a hundredweight; and the twentieth of a ton is
either 100 or 112 pounds. The quart and the gallon signify in
each case three different measures, and in the United States
there have been 130 different measures called bushel, none of
them conforming to the bushel of England.

In the vocabulary of the metric system there is one specific,
definite, appropriate word to denote the linear unit, one for
the unit of area, one for solid measure, one for the unit of
capacity, and one for the unit of weight. The word is exclusively applied to the thing, and the thing is exclusively
denoted by the word. Thus, the metre is a definite measure of

FORUM, August, 1901, vol. 31, 739-744.

length and nothing else; it has only one value at home, and can have none other abroad. This system employs five unit words and seven prefixes, or twelve words in all, while in the English system there are seventy-four units, having fifty-six names, eighteen of which are ambiguous.
. . . .

One of the strongest objections urged against the adoption of the metric system is the disturbing effect its introduction would have, for a time at least, on the daily vocations of the people. "How would the servant-girl," some ask, "know how many litres of milk to buy if in all her previous purchases she had bought only quarts?" Suppose she knew nothing of the change, and that it began during her sleeping-hours, on a certain date. On the next morning she would hail the milkman and say, "Let me have two quarts of milk." He would answer, "We don't have quarts any longer." "What do you have, then?" "Litres," would be the reply. It is not likely that she would be terrified by the word, but she would ask to be shown a litre measure, and to her eyes it would look so much like a quart that she would at once know how many of these units of milk she would need. The result would be similar if a lady should try to buy a certain number of yards of goods. On finding that she could not have the articles in yards she would ask to see how long the metre was, and she could easily form a fairly accurate opinion as to how many she would require. In the great majority of the petty transactions of our daily lives we are guided in our purchases by the amount we wish to expend, or by the size of the pile the shopkeeper measures out or puts in the balance.

The change in our system of weights and measures would be accompanied by considerable cost, for new scales or weights would be required. As, however, the average life of a counter scale is about two years only, the merchant, knowing in time when the change would be made, would not incur much expense by either hastening or delaying somewhat his purchase of a scale so as to buy one of the new system. Some new machinery would have to be constructed and new dies cut; but even at present all manufacturers who are seeking foreign markets must have machines which conform to the measures of the country where they are striving to create markets. Many of our shops already keep two sets of machinery on this very account. And it is idle to assume that our customary units would become entirely obsolete on the day of the adoption of the new system. The most radical proposition goes no farther than to say that, after a certain date, contracts, in order to be legal, must be drawn according to the new system. For years to come the use of the familiar units would be permitted, just as at present hundreds of units are in use every day that have no legal recognition whatever--for instance, the bit, the levy, and the hand.

. . . .

If any question should arise as to the accuracy of scales or
of any of the instruments used in determining length or capacity, such a question on final appeal must be referred to the
Government for decision. But the English standards are not employed in making the comparison. In the case of a question of
weight, neither the pound troy nor the pound avoirdupois is
called into service, but the kilogram. And although the Government possesses standard bushels and gallons, reference is
made to the international litre wherever an authoritative de-
termination is asked for as to the accuracy of a measure of
capacity. We therefore see that the adoption of the metric
system simply means its introduction into every-day life and
nothing more, as our customary units already have their values
given in terms of the metric units, and are determined by comparison with them. It is a mistake to suppose that it is the
scientists who are clamoring for the introduction of the metric
system. They have it and use it whenever they wish to have
their writings perfectly understood by the scholars of the
world. Apart from this reason they use this system in their
studies, because of the ease with which it enables them to as-
certain specific gravity and to express measures of capacity
in weight.

. . . .

At the present time we are seeking to enlarge our trade with
nations that use the metric system, or in countries where our
strongest competitors are nations using that system. The disadvantage in both cases are identical so far as concerns the
use of a system of weights and measures differing from that employed by our customers or by our competitors. The American
price-lists are unfamiliar, and the amiability of the prospective buyer must be drawn upon before attention can be paid to
our goods. Then, too, there is no easy standard of comparison
with the products offered by foreign competitors. The difference of monetary systems alone is a source of sufficient
trouble. When it is increased by the unlikeness of the units
of weight and measure, the problem of making a double conversion possesses difficulties for the would-be buyer equalled
only by our youthful perplexities in dealing with the "double
rule of three." Owing to the likelihood of making errors as
well as the trouble of making such conversions, our price-lists
and quotations make but little headway in the introduction of
our manufactures into foreign lands.

The adoption of the metric system by this country would undoubtedly aid us in trading with nations that already use it.
And if it would aid us in selling, it would also help us to

buy, by placing larger means at our disposal. Then our increased prosperity would be accompanied by greater prosperity for the other members in the family of nations, and the circle of exchange would be thereby enlarged.

THE CONDITION OF WAGE-EARNING WOMEN.

CLARE DE GRAFFENRIED.

OUR wage-earning women and children are grouped into five classes, New England being preëminently the textile district. Each of these States contains model industrial establishments supporting excellent tenements or separate homes for workers, libraries, clubs, even hospitals for the sick or the injured operatives, as at Lowell. Factory laws have reached their highest development in Massachusetts, other New England commonwealths lagging behind. There the sixty-hour week prevails, reduced to fifty-eight in Fall River; children under thirteen are forbidden in mills or shops, and for those under fourteen twenty-two weeks' schooling each year is obligatory; accidents must be reported and employers are liable for damages. Thirty-three inspectors, two of whom are women, enforce these acts.

Manufacturing New England, Puritan and homogeneous no longer, speaks a French *patois*. The influx of French-Canadians and their domestication and economic effect on wages began with the stimulus given to manufactures by the Civil War, when these operatives were brought and housed in hastily-constructed tenements, without drainage or conveniences, which still exist around many mills, a reproach to civilization. Setting up a tiny stove and spreading mattresses on bare floors, parents and offspring overran the factories, hoarding wages for a time, then returning home to buy a farm. Illiterate, living wretchedly, inspiring fear and aversion in neighbors, unfamiliar with the English tongue, they overran entire districts, and there are to-day "little Canadas" of crowded barracks owned perhaps by fellow-Canadians become investors, each house populated by from three to fifty families, who dwell under social conditions truly amazing. Eight married couples, besides six tenants, board in a tenement of three or four rooms, women sleeping in one apartment, men in another, their children left in Canada or farmed to aged care-takers. Only two of these adults can read, but all own bank-books and are industrious.

Higher-grade and domesticated Canadians purchase homes, support churches and parochial schools, French newspapers and social clubs; and they buy abandoned New England farms. Aggregate family

THE FORUM, March 1893, 68-76.

earnings reach the thousands, for fifteen pledges of affection surround many boards and twenty-one scions of the same parents is no rarity. Wherever this prolific race appears wages drop. With lower standards of living at first, and later with superior thrift, they underbid Nova Scotians, Irish, and English, becoming as formidable rivals as our bread-winners' other foe, the childless married worker—New Englanders being so often childless as to encourage divorce and threaten the perpetuity of the race.

Boston toilers winning industrial prizes share the city's intellectual life. Suspender-makers belong to Browning clubs, a necktie-worker writes magazine verses, a "button-holer" buys oratorio tickets. More than other wage-earners, Boston girls are homeless, dwelling in lodgings and eating at restaurants, neglected when ill, without resources if unemployed; their path shadowed by competitors eager to displace them at less pay—Italians or high-school graduates working for pocket-money. Merchants advertise for "girls who live at home, to learn the business"—a device to fill the firm's coffers by inveigling artisans and small shopkeepers into presenting their daughters with board that the daughters may engage for wages which will barely clothe them. This unwise, uneconomic gift from parents enables the daughters to set prices for adults at a figure on which no unassisted girl can live. Were subsistence charged for at home, the worker must earn more and spend less. Mechanics and clerks pay from their own pockets half the wages due their daughters from establishments such as these. Of 17,427 wage-earners, one-tenth earn but one hundred dollars to one hundred and fifty dollars yearly, less than two dollars and three dollars a week—a pittance that scarcely buys food. Truly self-supporting females are often in rags and broken boots, destitute of over-shoes, umbrella, or waterproof, and without flannel in winter.

Working mothers are equally wretched, spending for their own needs not five dollars a year, debts and insurance absorbing gains and the husband's ninety cents to one dollar and a half a day being insufficient to feed eight mouths. With envy such a wife describes "two old maids" who have laid by "a b'ilerful o' money." For herself, until her "young ones" labor there is no outlook, only fear that "Frenchy in the next alley" may take away her six looms.

Nothing being saved toward marriage, newly-wedded couples buy furniture on usurious instalment plans, the bride weaving to meet the payments. Children come fast, illness and death occur, grocer and landlord clamor; furthermore, loss of employment befalls even prudent, industrious husbands. Then the already overtasked mother leaves one child in charge at home and returns to the mill. What happens? In a hovel one day I saw a maid of nine with five younger dirty toddlers in line before her, weeping loudly. She was beating each in turn with a heavy leather strap. "They needs it," explained the tiny guardian. "Ma, she hollers at 'em, but I ain't got no big voice like ma, so I lets go at 'em with the strop three times a day."

Native New England workers are not extinct, and on them devolves the care of parents and dependents. That industrial men support the aged, orphaned, and helpless is a popular fallacy. Since women have gained independent livelihoods, men shirk family ties, throw off their parents, marry rashly, and often disappear when an heir is born, most mothers of *crèche* children being deserted wives.

The second group of working-women focuses at New York, with environment wholly urban, a congested industrial population complicating the economic problem. Bad factory or home surroundings intensify the life-struggle, offsetting higher wages, factory supervision, and just and liberal terms of employment. Here reckless competition and middle-men cut prices; here immigrants with low standards and of temper alien to American character submerge all callings and "freeze out" domesticated workers who contribute to our national character traits and capabilities which strengthen our civilization. Here mercantile houses offer as wide a field for abuses as for advancement, and unscrupulous dealers cheat, oppress, and decamp without paying employees. Finely equipped, admirably conducted factories and shops in every industry are flanked by hideous rookeries.

Of late, excellent factory laws, enforced by ten men and eight women inspectors, have improved shop conditions in New York and Brooklyn, reducing child-labor fifty per cent. Children under fourteen are excluded from manufactories, also those under sixteen who cannot read and write English. Though education is compulsory, insufficient school sittings reduce the benefits of this provision. The present statutes are defective, too, in omitting mercantile firms from responsibility to the law and in failing to protect adults as well as minors by the ten-hour clause, since not otherwise can sweaters be prevented from driving hands sixteen and nineteen hours a day during many successive weeks, seven days in the week at that.

Native workers of native parentage are scarcely found, the best of this class having risen above manual into semi-professional vocations,

44

and those less plucky and ambitious having yielded to our assimilated immigrants in qualities which command success—endurance, deter-mination, adaptability, thrift, and energy. Certainly the American native for two generations is not found at the top. Forewomen, head dressmakers, drapers, buyers, higher-paid saleswomen, are either of foreign parentage or outright foreigners. The native is found often at the bottom. Unmarried or widowed, gaunt, hollow-eyed, bitter against the hordes who, subsisting on scraps and working for a song, rob her of bread, she still bristles with prejudices, will neither consort with Catholics nor be "bossed" by Jews. Unable to pay board by a saleswoman's wage on which her Hebrew neighbor thrives, she drifts into a cloak-shop, and stranded in dull season, finishes knee-breeches in a garret for six cents a dozen. No friendly working-girls' club can lure her, no industrial millennium will dawn for her, and her sister with less pride and stamina sometimes sinks into pauperism or prosti-tution—where Jews are rarely seen.

Two types of wage-earning capacity congregate in cities—the ablest, busiest, most fortunate, and numerically the stronger; and the neediest, idlest, most incompetent, who hang on the outskirts of trade from want of energy to penetrate it or from inadaptability to new surroundings. All remunerative pursuits attract recruits who in the absence of trades-unions undermine wages by competing at lower rates. Wherever the rewards of industry are greatest competition is fiercest, and no woman's place or stipend is safe or fixed, each individual being but another atom flattened by livelier protoplasm.

True, in every metropolis many girls earn good wages, live com-fortably, save, and dress well. But consider home conditions where two million souls inhabit tenements! The belongings of prosperous bread-winners mould in wretched, unlighted rooms of buildings eight stories high and twenty feet broad, inhabited by from thirty to seventy families, with a rear tenement three yards away and a third huge hive behind, swarming with people and reeking from neglected vaults. Women who must climb home over the bodies of drunken tramps asleep on the filthy stairs take little pleasure in a piano whose music is drowned by the orgies of fighting neighbors. Mattresses, by day strapped to doors or hung out of windows, at night cover tenement floors to pack in hundreds of men and girls—illegal lodgers. By the census of 1880, more than eighty-two per cent of New York house-holds resided in dwellings containing three or more families. Recent comparison by the Massachusetts Bureau of Statistics of assessed value and rentals of six hundred and forty-six houses in bad sanitary condition in Boston proves that the incomes from these death-traps average twelve per cent on assessed valuation, reaching in many cases thirty per cent and even forty-nine per cent. In New York, where the least habitable dwellings are the most profitable to landlords, Trinity Church itself owns unsanitary tenements without water above the second floor, and recently won on a technicality a suit brought by the Board of Health to compel the corporation to place water on other floors.

In the poorest quarters of foreign cities—Whitechapel in London, Glasgow, Rouen, Naples—I saw no such concentration of squalor, filth, airless dark pest-holes, foul closets and sinks, such massing of human beings within narrow, damp, windowless walls as exist in New York tenements. And I busy myself only with workers—women and children following a recognized occupation while employment offers or health permits—not with charity cases, the sick, the unemployed, the criminal classes. I "slum" because wage-earners must live in the slums—not always full-time workers, however, since laborers at low pay not only get odds and ends of work, but have also such debased standards of comfort and so few wants above animal wants that three or four days' gains keep them a week in their miserable fashion. Such desultory intermittent bread-winners supersede the industrious, under competitive methods, by hiring at starvation rates and levelling all pay to their wretched requirements. The moral points itself: increase standards of comfort by educating and uplifting the masses and, so far as economic laws allow, wages will rise.

The third industrial group sweeps westward, Philadelphia leading in population and in the number of shop-workers without male providers and therefore self-supporting. Throughout Pennsylvania, in Baltimore, Buffalo, Cleveland, Indianapolis, and Louisville, while thousands of female wage-earners must work or starve, other producers are partly maintained by iron and railway employees, mechanics and skilled laborers, for whose womankind daily toil is not indispensable, although the wife's or daughter's gains change a precarious into a comfortable family existence.

Notwithstanding economic supremacy, Pennsylvania has no factory code. In 1889 an act was passed restricting the labor of minors to sixty hours a week, excluding children under twelve from factories and stores, providing for the safety of women and juveniles, and appointing five inspectors. But unscrupulous employers may and do work minors twelve to sixteen hours daily if a mill breaks down, and the hours that adults are required to toil are unlimited. Loathsome tenement factories hiring less than ten persons are free from inspection, and payment of wages semi-monthly is evaded. Maryland's ten-hour law applies only to minors.

Though the tenement nuisance taints every city—thirty-three per cent of these structures in prosperous Buffalo being highly unsanitary —all this region is the paradise of separate homes. In 1880 less than five per cent of Philadelphia houses held each three families or more, and eighty-four per cent were occupied by single families. Such social surroundings mitigate the lot of woman-drudges and child-slaves; but struggle for self-maintenance always means hardship for this weaker economic element. Native workers preponderate, mostly country girls, a large percentage of whom lodge and board. The moral tendencies of bread-winners here and in Baltimore are high, many toilers teaching in Sunday-schools and zealously helping their churches, practical Christians "giving to the Lord" the literal tenth of scant incomes. A mother and daughter in Charleston with combined wages of five dollars a week set aside fifty cents weekly for the church before buying food, and a Fall River weaver with four to support on three hundred dollars a year gave always his tenth. Catholics are fully as liberal.

While the great textile mills of Philadelphia are unsurpassed, other mills and factories there are unsanitary and without separate closet accommodations for women. Much woollen yarn is made by such juvenile labor that the product is bad and causes woollen and carpet weavers elsewhere to be fined. Cotton factories around Baltimore provide good homes with gardens, libraries, and schools; illiteracy, nevertheless, is still rife, for here and westward first appear the "poor whites," as unlettered as peasant immigrants. Factory supervision not existing, children under twelve labor illegally; four-year-olds shell pease and stem strawberries in dirty canning factories among six hundred polyglots from five A.M. till night. At work in tobacco little ones displace the adults, and glass-manufactories notoriously violate New Jersey and Ohio child-labor laws, the strictest yet framed.

At work on clothing and underwear wages have fallen literally to a starvation point, a fact incontrovertible, although families half-supported by father and sons grow rich at such pursuits. Factories turning out underwear and overalls so specialize employment that the strain shatters the operators' health, in spite sometimes of shop comforts. In Quaker villages and Bohemian and Polish colonies, partly maintained home-workers make a dozen garments for the price due needy seamstresses for one. Men's shirts at forty-nine and thirty-seven cents a dozen, men's drawers at twenty-seven cents a dozen, and thread, machine-rent, and expressage deducted—these rates, accepted for pocket-money, become the highest wages that can be obtained by immobile self-supporting labor excluded from factories because of invalid or infant dependents. "Wholesalers" abandon manufacturing to escape wear and tear, rent, and changing patents in machinery, preferring profits through the sweater, though professing to desire organization among needle-workers to resist present killing rates. This organization—useless if it fail to comprise the homes and tenements—is unlikely to occur. To further it is one of the holiest missions of the century.

The Southern wage-earning group presents unique conditions. New occupations—any occupations indeed—for women are a boon to this impoverished region. Breaking down old traditions that labor is menial and degrading, promising industrial independence, providing employment alike for Negroes, poor whites, farmers' daughters, middle-class families, and needy ladies, paying trades and gainful pursuits in the New South bring a righteous emancipation. The home-labor supply being unlimited, cotton, coal, and iron being cheap, and the cost of living being lessened by a mild climate, industries grow apace; and notwithstanding the inertia characteristic of agricultural countries and increased by slavery, the centre of gravity of the population is shifting, cities are absorbing population from the rural districts, small tenants are deserting their holdings to crowd the homeless and destitute into manufacturing centres, where some become good operatives and acquire a competence, others work fitfully, and many live by the toil of very young children. Ownership of homes has ensued to a certain extent, but there has been a greater stimulus to corporation houses and villages, some excellent, some sadly disgraceful. Nowhere is education compulsory; the public schools, new and ill-provided, fail to reach vast numbers even of the white children, and illiterate poor whites neglect frequently the teaching provided, while the blacks monopolize it. Mississippi and Georgia, leading where Northern States have not followed, have established admirable industrial and trade schools—*not* reformatories—for needy white girls. The weekly working hours in the South average seventy-two, and to secure half-time Saturday, children and adults labor thirteen and fourteen hours daily in mills and stores. Georgia, more advanced than New York or Pennsylvania, limits toil for adults to sixty-six hours a week; but children seven and eight years old are permitted to drudge eleven hours a day. Virginia's law, often violated, excludes children under twelve. Louisiana possesses a fine factory code and the ten-hour day, without any enforcing power except the police. In other States the straggling enactments recorded are frequently disregarded, State control being practically a dead letter. Consideration for women is always shown, and the best employers shorten the working day and pay their operatives weekly.

The poor white, bare of comforts and often ill-paid, has in the Negro an industrial rival who easily subsists on less and underbids her at laundries, tobacco-factories, and in sewing. Higher-grade Southerners seek employment, even against the family wish, at clerking, dressmaking, patent medicines, binderies, textiles, box and cigarette factories, free from the blacks' competition, earning pay enough to lessen the pinch of poverty. Tobacco affords a lucrative pursuit without odium, darkies appearing only in its lowest branches; and in a model Richmond factory the cigarette-girls equal the best Northern workers in position, manners, and education. Nowhere else in the world do so many well-bred women, bankrupt and bereft of male providers, labor at manual callings as at the South, pursuing without loss of caste vocations which elsewhere involve social ostracism.

The last gainful division belongs to the Northwest and half-European cities like Cincinnati, St. Louis, and Chicago, part of whose foreign protelariat pursues un-American ideals of comfort and morals. As in New York, conditions characteristic of life in all big cities appear to be, and yet are not, disadvantages inherent in labor itself: tenement abominations, high rents, the "submerged tenth's" importunate rivalry depressing wages and debasing ethical standards, the rush for gain trampling out justice and humanity. Sweaters and contractors swarm; tenement workers delve far into the night, sometimes alongside the corpse of some poor baby that died of inanition. Separate homes are less infrequent, air and light more plentiful than in New York; yet many streets consist of ill-constructed houses holding each from three to eight families—a population in degradation.

No factory laws exist in Illinois, but Chicago has municipal regulations enforced by inspectors of both sexes. Missouri mills run twelve hours a day. Such restrictive statutes as Western States have passed, being for the most part unaccompanied by supervision, present few obstacles to hard or unprincipled employers. All-night work is sometimes required of women, and juveniles of tender age are employed in many industries. A recent Ohio enactment empowering inspectors to abolish child-labor in dangerous and unhealthful spots has done much to root out this evil. Natives with difficulty hold the best places against the higher-class immigrants who are the back-bone of the Northwest's wonderful development; but the lower working class, coming of pauper stock in the Old World, remains illiterate and untrained, without attachment to our soil or institutions, demoralized by sudden liberty and hostile to our labor ideals. Even its Americanized offspring, after the sterile, misdirected teaching of public schools, are difficult to manage, and employers have recourse to harsh and unscrupulous methods. Great is the need for skilled labor. Incompetent, exacting, unreliable is that which offers; hence are bred antagonisms such as the kindest and justest proprietors cannot dispel. Capable workers suffer from fines and restrictions aimed at the careless and untrustworthy; "docking" for lost time creates discontent, and petty, tyrannical rules diminish the toiler's self-respect. While wages are better than in the East, living is dearer. From the partial adoption and abuse of foreign customs, Sunday dance-halls and theatres, with bars attached, attract the rougher work-girl. Numbers of virtuous bread-winners, however, contribute regularly to the family support, work under just and liberal employment, live in fairly good homes, and in due time marry. But the same class that suffers in the East suffers more in Western cities from the sharper, unworn wheels of the new industrial Juggernaut....

A FORGOTTEN INDUSTRIAL EXPERIMENT.

By Sara A. Underwood.

NE of the outlying suburbs of the city of Springfield, Massachusetts, constituting the eighth ward of that city, is a small manufacturing village called by the somewhat romantic name of Indian Orchard. About forty years ago this was the scene of an interesting experiment in social industry, an account of which may be of interest in these days of social settlements and of wide consideration of industrial problems.

In 1854, on a sandy flat in the hollow of a slight eminence about one mile from the station of the Boston and Albany railroad, the village of Indian Orchard consisted of one big brick cotton factory, two huge, dreary looking brick boarding blocks, divided into eight boarding houses. On higher ground on the hill was a short row of cottages with pretty yards in front, known as the "Overseers' blocks" near which stood the modest Congregational church and a small schoolhouse. Half a dozen straggling private houses, in one of which a milliner had rooms, and a little wooden structure near the "Jenksville road," which combined the accommodations of a country store and village post office, made up the whole of Indian Orchard, the real life of which centered in that treeless, flowerless, sand plot and its bare looking new brick buildings devoid of the slightest attempt at architectural or landscape adornment or beauty.

Though to a stranger's eye the place thus seemed extremely ugly, dreary, and isolated, yet "the Orchard" held many nooks and walks full of interest and beauty. The Chicopee river flowing directly past the back of the factory wound through the village by lovely banks clothed in verdurous beauty, onward towards Chicopee. At a short distance from the boarding houses, near the entrance to some wooded spaces in those days, a clear, pretty spring of mineral water welled up cool and sparkling under gracefully bending birch trees; and in every direction bits of woodland offered inviting walks and a host of wild flowers for the picking during the season.

But it was not these partly hidden attractions which brought to the dreary looking sand plot and its unlovely bare brick buildings, brought from near and far to that isolated little village, a more ideal set of employés perhaps, than has since been gathered together as factory help in any place in the United States. How, or by what means, or by whom planned, the writer of this knows not, but is given to understand that the "Indian Orchard Company" was started about the time designated as a sort of working-girl's Utopia. Its business headquarters were in Boston, and its first overseers and employés came from Lowell. As it was started not long after the collapse of Brook Farm, and less than a score of years after Dickens's glowing description of the life of the Lowell factory girls had appeared in his "American Notes" (a description then no longer true of Lowell, by reason of the influx of foreign workers), and at a time when "transcendental" ideas were still in the air, it must have been an effort of some enthusiastic humanitarians to carry out an ideal industrial scheme on a somewhat restricted basis.

For it was mainly American women and girls who were to be benefited in this scheme. Although at that time in both cotton and woolen mills in

NEW ENGLAND MAGAZINE: AN ILLUSTRATED MONTHLY, July, 1898, vol. 18, 537-542.

New England male help and foreigners of both sexes were employed in all departments indiscriminately, it was soon understood that in the new Indian Orchard mill no foreigners of either sex would be given employment, and that women only were to be employed in departments hitherto monopolized by men, such as the spinning and dressing rooms, as well as in the weaving rooms, and that the fewest men consistent with the needs of the plant would be employed by the company. The male help were mainly overseers and their assistants, with machinists and others connected with the work. When, a little later, a gardener was needed to cultivate flowers around the mill yards, a man was also employed for that. No children of either sex were employed.

It was also understood that the relations between the overseers (at that time a dominating, often hated class) and the help they employed were to be of the friendliest character and that the mysterious "Company" would discharge any overseer or assistant who used harsh or offensive language or methods towards the employés. The Company's "Rules" were hung up in conspicuous places in all the rooms. The duties of both overseers and the help were therein plainly stated, and whoever infringed those rules was to be quietly discharged from the Company's employ. Thus it was against the rules for any employé to sew, crochet, or read books or newspapers during working hours, because they were liable to become engrossed by these to the detriment of the work on which they were engaged. If some smart girl, presuming on supposed favoritism, transgressed that rule and was observed, as she was likely to be, the overseer, instead of scolding her personally, as in other manufactories, would post a notice where all general notices were posted, to the purport that it had been observed that one or two in the room were breaking the rule, and such were hereby warned that a second offence would cause their dismissal. If the warning was not heeded, the offender would be given a note informing her that the wages due her up to date could be obtained by an order at the desk, as she was no longer in the employ of the Company.

Though the wages paid were not remarkably high, they were slightly above the average in other cotton mills, while competition and attention to the quality of the work done were stimulated by small money awards given to those whose work was above the average in perfection and amount; also trifling fines were imposed when glaringly faulty work was found. A larger number of extra hands were employed in this than in other factories, so that the employés could oftener take a day off, of course with loss of the day's pay; also the rules were less strict in regard to permission being given for an hour's absence without loss of pay. Indeed, so friendly were the relations between employer and employé, that it was not an unknown occurrence, when help happened to be scarce, for the overseer or his subordinates to take the worker's place for a short time when the need seemed great, out of pure kindness. One such occasion was related of a May morning when a Vermont farmer's pretty daughter and her friend, desperately longing for some of the trailing arbutus which they knew was in bloom on the woodsy "Jenksville road," though there was no extra help that morning, ventured to make an honest statement to their overseer of their longing for the fragrant blossoms, asking permission to take a run in that direction, promising to be back as soon as possible. He listened sympathetically, assented to their request and, calling to an assistant overseer, asked him to take the place of one of the girls, while he himself took that of the other. No doubt they both felt themselves sufficiently rewarded when a little later the two girls returned with roses on their cheeks, a brighter light in their eyes, and their hands filled with fragrant Mayflowers, of which a fine bouquet was the reward of the kindly "bosses." This was not an unusual incident, but was quoted to the writer merely as an illustrative one.

The boarding houses were furnished in every department by the "Company"; thus the boarding-house keepers suffered no loss or extra expense in taking charge, while a change of keepers caused no special disturbance to the boarders. The employés paid a less amount for board than in other manufacturing villages, as the company paid the keepers an additional percentage for each boarder; also, to guard against any loss through dishonest boarders, an amount equal to the board due was kept back from their wages when thought necessary. The private rooms of the girl boarders were spacious and airy enough for two, the usual number who occupied one room, though on each floor one or two extra large rooms were occasionally occupied by four girl friends. All the rooms were furnished plainly but comfortably, with good beds, washstands with bowls and pitchers, plain chairs, and curtains; there were no carpets, but painted floors. Not infrequently a less healthful but in some respects a more agreeable arrangement was allowed to be made by the girls themselves, when four friends occupying two rooms turned one into a double-bedded room, keeping the other as a private sitting room, bringing from their own homes a carpet, rockers, etc.; and buying from their own funds a small stove and fuel. Here they could gather around them their special "chums" among the boarders, for a quiet chat, cheap refection, or to sew or read. There was, however, on the ground floor of the boarding houses a barely furnished reception room, while the big dining room, whose two long tables did not fill half its space, served also as a common meeting ground for all the boarders.

As the Indian Orchard mill with its superior class of help was for a number of years so well known that it became much of a "show place," attracting visitors from all parts of the country, the girls themselves, many of them really beautiful and fairly educated, affected a more attractive dress than was usual among factory operatives, though no distinctive attire was attempted. This would have been generally resented, since a strong individuality was maintained by all, but white collars, pretty brooches, and nicely laundered, well made, neat-figured print dresses were generally worn by all, and black silk aprons, embroidered, braided, and tied with handsome silk tassels, were worn by the majority of the younger girls in going to and from the boarding houses to the factory. These were such a short distance apart that in summer time often only a barege veil thrown carelessly over the head was the only covering worn — or a parasol was carried. When on Sundays, in deference to an expressed expectation in the Company's rules, the employés attended church, their pretty toilets were in touch with the latest fashions of that day.

How the news of the new industrial experiment at Indian Orchard was first circulated and reached those who first obtained employment there I do not know. It is probable that, as the first comers were selected from the Lowell mills, and in those days most of the New England manufactories employed a considerable number of well-educated American women, those first employed spread the news of the ideal mill in letters to their farm homes and to girl friends working in other places. At any rate the good news spread, and many girls who had not hitherto "worked for a living," though they had longed to earn something for themselves, gladly offered their services. Of course many failed to obtain places. Certain it is, however, that a fine class of help was secured. The majority were farmers' daughters from Maine, New Hampshire, New York, Vermont, as well as Massachusetts. Some sought work for the purpose of paying off the mortgage on the home farm, others to get money to fit themselves as teachers; still others had been teachers, but hoped to earn more in the mill. There were also a number of "old maids," deserted wives, and young widows with one or more children to support.

Though here as in all places the congenial minded formed themselves into groups of special intimacy, there was yet a most pronounced spirit of *camaraderie* about all belonging to "the dear old Orchard," as well between the overseers and help, as among the employés, and a new girl, however unknown, was sure to be met with kindly looks and friendly smiles from all "the old hands" as an antidote to any possible homesickness from which she might be suffering, and a few days only would suffice to make her one of some congenial group. Here friendships were formed which lasted a lifetime.

Two years or so after the Indian Orchard experiment was well under way, a new interest was added in the gift from the Company of a library and reading room, which led to the formation of a lyceum, in which the girls took a leading part in the debates and entertainments. A manuscript paper was also started, filled with original contributions from the employés, in prose and verse, some of which would have done credit to periodicals of more pretensions. Its first editor, elected by secret ballot, was a modest girl of twenty years, one of the weaving-room girls. Most of those who succeeded her later were also young women employés.

The library was stocked in a manner characteristic of the philanthropists behind the scheme. Two thousand dollars were devoted to the purchase of books, the greater part of which were selected by the donors. Then a notice was put up in each of the workrooms asking the employés

to send a list of such books as they personally would like to have added to the library, and if the committee in charge decided such to be admissible they would be bought from the fund reserved for that purpose. Some of the girls sent in quite long lists, and most of the books thus selected were duly procured and added to the library. That the pervading spirit of the "Transcendental school" was that in which the Indian Orchard scheme originated, was plainly indicated by the trend of much of the reading matter selected by the Boston committee. Among the books were Emerson's Essays and "Representative Men," Theodore Parker's Speeches and Sermons, "Memoirs of Margaret Fuller," by Emerson, James Freeman Clarke, and Channing, Thomas Cárlyle's Essays, Mrs. Browning's "Aurora Leigh" and other works of like character. The earlier poems of Lowell, Longfellow, Whittier, and the Cary sisters found place, and the better class of fiction was represented. In the reading room the *Atlantic Monthly*, then in its first youth, was a welcome visitor and we may suppose a helpful force in life to many.

Talked of as a singular industrial experiment, Indian Orchard, isolated as it seemed, had many visitors in those earlier years. The different floors of the mill were airy and well lighted, the machinery painted in light, pleasing colors, the broad, clean window ledges made attractive with clean white curtains and many potted plants in bloom. The smooth painted floors were kept perfectly free from dust and litter, being swept four times a day at stated hours, each employé sweeping quickly over her own area, then passing the brush to her next neighbor, while scrub-women were employed to scrub all the floors twice a week. A section of one of the Company's rules read as follows: "All persons employed as help must be particularly attentive to the cleanliness of the machinery upon which they work, and not forgetful of neatness in their own personal habits;" therefore the handsome machinery was kept in "spick and span" order, while each girl took special pride in the prettiness of her own apparel. As many of them were bright, keen-witted, lively girls, whose intelligence was clear to the most obtuse observer, the visitors were given much food for thought.

Among the improvements upon the earlier methods of factory life introduced in this experiment was that of adding roomy washrooms as adjuncts of beforetime limited toilet rooms. In these could be found a good supply of soap and hot and cold water, with plenty of sink space for all. Beds of bright flowers made the outside yards as attractive as was all within doors.

As nearly all the workers were far from their own homes, boarding in the two big brick blocks — the city of which the village formed a part being seven miles away and the road between showing only a few scattered farm houses — of course the village became a little community by itself. Its amusements were not many nor varied. About every boarding house had at least one inmate who owned or rented a piano, which held a place of honor in the general reception room. A few old-fashioned melodeons accompanied their owners to the Orchard and found place in a corner of their rooms; others owned accordions; one or two violins and cornets were owned by the male boarders; so there was considerable noisy blowing and strumming of instruments of an evening, some girlish singing in the rooms, once in a while an impromptu dance in the reception or dining room; but these were not kept up to a late hour, as the rules were imperative in regard to closing the doors of the boarding-places at ten o'clock. In the winter there were occasional sleigh rides and some coasting, while on Sundays going to church, riding, walking and writing letters were the order of the day. But generally the days passed on without much variety. Most of the girls were their own seamstresses and dressmakers, and the evenings were occupied with sewing, reading and writing letters. Not infrequently little coteries of friends gathered in some girl's room to chat and enjoy together some little treat of fruit or something more substantial sent from the home farm, while they gossiped of "Orchard" events, about their homes, or about some book. Most of the girls took vacations once or twice a year to visit their homes and old friends. These vacations were the chief events of their lives in those days, and oftentimes a homeless friend was invited to share in these vacations, to make new friendships in other girls' homes. Frequently during the hot weather in summer the overseers treated the help in their own departments to pails of cold lemonade paid for out of their own pockets.

The great financial crisis of 1857 was one of the first of several disintegrating forces which served eventually to break up the ideal state of things at Indian Orchard. By the autumn of that year, panic ruled every phase of business; banks were breaking, credit was not to be had, all manufacturing business came to a standstill, and the cotton mills of the country were especially affected. George W. Holt, who had come from Lowell at the beginning of this industrial experiment as overseer of the weaving department, and whose genial manners and cordial appreciation of the spirit of the scheme had made him a general favorite as well with the Company as with the help, had been promoted about this time to be agent in charge of the mill. His entrance into any of the rooms was usually the signal for warm greetings and smiling looks. But one cloudy afternoon in the autumn of that year the girls of the upper weaving room were surprised to see him enter the door with gloomy face and averted eyes; and, passing into the overseer's private office he was soon followed by the overseers of the different departments, apparently for a business consultation of some sort. Each overseer as he filed out again with a slip of paper in his hand looked worried and anxious. Mr. Holt himself tacked up in this room, where the employés were especially attached to their old overseer, one of these slips, and, pulling his hat over his eyes, passed out quickly, exchanging salutations with no one. In a few minutes all the girls knew the purport of that notice. It was addressed to the employés generally, and was to the effect that orders had been received from the Company's headquarters in Boston that in consequence of the prevailing financial panic the mill must be shut down as soon as the small supply of cotton on hand was used up. This, though a business notice, was couched in words of fraternal kindness and of heartfelt sorrow over the possible breaking of the pleasant ties which subsisted between the employers and employés, with strong expressions of a hope that it might be only a short time before the amicable relations would be resumed. This was signed by the agent himself, as a personal document.

A heart-breaking spectacle ensued. Every loom in the room was stopped as by one impulse, and nearly every girl was weeping, gathering together in little groups to discuss the outlook, the largest group surrounding the overseers, who looked grave and sorrowful as they explained what they understood of the situation. Six weeks afterwards the last looms stood empty, and the factory was "shut down" for an indefinite time. But the humanitarian spirit which suggested and carried on the enterprise controlled the action even through this shutting down process. Those girls who had homes near by to return to were the first to be discharged; those whose homes were farther away went next, with a little more money on hand to get home with; the widows who had children dependent on them, and homeless girls, were kept to the end of the work.

It was not until a year later, December, 1858, that the Indian Orchard mill started again. Word was sent of the re-opening to all the former employés whose addresses could be obtained, and a majority of these returned, among them many who had found work in other places but left it to return to the Orchard mill. In 1859 business was again flourishing with the Indian Orchard Company, and a new mill and boarding houses were erected not far from the old ones. With the breaking out of the war in the spring of 1861, when the financial state of the country was in consequence somewhat disturbed, and through other causes, one of which doubtless was the fact that the former ideal arrangements did not prove so successful from a business point of view as had been hoped, a different course began to be adopted. Agent Holt fell ill, and a new man, a stranger, was put in his place. Foreign help, especially French Canadian, began to be employed. Wages were paid according to the scale of other manufacturing plants of the same kind, and deterioration began. Later, different kinds of manufactories were established in the village, which still further changed the moral and intellectual characteristics of the population. The ideal standard was lowered, and in a few years Indian Orchard became like other New England factory villages.

From a worldly point of view it must be conceded that this noble attempt to elevate the condition of the factory workers in New England was far from a success; yet the sweet spirit of fraternity, of moral and intellectual progress was there engendered and illustrated in such a way that, though no great or definite results can at this date be pointed out, yet sure it is that the world into which those earlier Indian Orchard employés went forth and spread their influence, was better because of the impressions which that experiment made upon their minds and hearts.

CONTEMPORARY PERSPECTIVE:
THE DISORDERED TECHNOLOGY

Is Technology the Cause Of the World's Problems?

ROY V. HUGHSON

It is an oft-made claim these days that technology is the cause of most of the world's problems. The corollary is that engineers are to blame, too.

But are we? Do the evils of technology outweigh the good derived from it? And, could anyone have foreseen the problems before they occurred?

This last is an important point. Most technology comes into being in an effort to make things better for people. The harm that it may do is generally unforeseeable at the time the technology is introduced, or arises later from abuses.

The difficulties in foreseeing problems are among the hardest that ever arise. Twenty years ago, nobody foresaw the effect that computers would have on our society. When the telephone was introduced, who guessed that it would change the courtship patterns of the country's youth? At best, an astute observer might have guessed at some of the changes it has brought about in the conduct of business.

The air conditioner was a boon to the worker who had to spend hot summers in a city office building. Since open windows lead to inefficiencies in the use of air conditioners, architects built structures with windows that didn't open. Such buildings were also cheaper, and stayed cleaner. But without air conditioning, these offices are unbearable in summer. So the electric utilities are forced to increase generating capacity to meet the air-conditioning demand. This inevitably results in more thermal pollution, and probably in greater air pollution. All because we tried to better the worker's standard of living.

Too, in this case, the good may outweigh the harm. Although people with lung disorders are harmed by air pollution, those with heart ailments are generally helped by a cool environment. And certainly the working public in general is better off in pleasant working conditions.

The Case Against the Automobile

Let's look more closely at the automobile—the machine that has probably caused more social change than any other technological innovation. Presently it is being blamed for two things. First, the air pollution that it causes; second, the number of deaths that occur through automotive accidents.

It has been repeatedly pointed out that the automobile has caused more deaths in the U.S. than have occurred in all the wars in U.S. history. And yet, when the automobile was introduced, it was hailed as having solved the problem of accidents caused by the runaway horse!

Don't laugh. The runaway horse was a serious accident problem in the horse-and-buggy days. If you've ever read any of the Horatio Alger type stories, you will remember the hero that started on his way to fame and fortune by stopping a runaway horse (which seemed always to have been pulling the carriage containing the wife or daughter of a rich merchant). This was a good fictional situation because the runaway happened so often in real life.

And, of course, nobody really imagined the speeds of modern cars. I remember a Tom Swift story in which the boy genius invented an auto that travelled at the fantastic speed of a mile a minute! Nowadays, few drivers are content with so slow a speed (60 mph.) for highway travel.

There is no question that some of the pollution caused by automobiles could have been (and actually was) predicted. These pollutants include carbon monoxide and nitrogen oxides. But could anyone have predicted photochemical smog? Even today, it's very difficult to investigate this phenomenon in the laboratory because interaction of the chemicals with the container itself ("wall effects") interferes with the experiments.

But at the time of the introduction of the automobile, nobody could have predicted these things because nobody would have guessed that the U.S. would ever have 50 million cars and trucks. One automobile isn't a problem. Even a million aren't really a problem. But 50 million are.

One prediction from the July, 1899, issue of *Scientific American* has been quoted in a recent book.*

"The improvement in city conditions by the general adoption of the motor car can hardly be overestimated. Streets clean, dustless and odorless, with light rubber-tired vehicles moving swiftly and noiselessly over their smooth expanse, would eliminate a greater part of the nervousness, distraction, and strain of modern metropolitan life."

I'm sure that the prediction made very good sense back in 1899.

The Problem Is People

With so many cars on the road, one easy way of cutting down on pollution problems is to cut down on the number of cars being used. One car carrying five people will produce only 20% or so of the pollution caused by five cars, each carrying only a driver.

This concept was tried recently in San Francisco (though the sponsors were trying more to cut down traffic than to cut down pollution). According to the *New York Times* for May 18, 1970, the operators of the San Francisco-Oakland Bay Bridge handed out 12,000 postcard questionnaires to commuters on their way from Oakland to San Francisco. The questionnaire asked whether the driver was willing to take part in a car pool. About 1,200 (10%) of the cards were returned. A computer was used to match people in a vicinity, and lists were sent out of people willing to join pools.

What was the result of all this activity? The bridge operators have learned of a total of *eight* car pools that have been formed.

Obviously, asking people to give up their automobiles just won't work. Indeed, the use of the family car is so much a part of American culture today that it could not exist without such transport. The American suburbs could not exist without the automobile (or its equivalent). And living *is* better in the suburbs than in the cities, which is why so many of us spend long hours in commuting in order to gain this advantage.

* Dubos, Rene, "Reason Awake: Science for Man," Columbia University Press, 1970.

The Insecticide Riddle

Perhaps closer to the chemical process industries is the problem of persistent chlorinated hydrocarbon insecticides. DDT is the main culprit, of course.

It took some time after the introduction of DDT before it was realized that it was taken up and stored in the fat of animals (humans included). It took even longer before the injurious effects of this accumulation were discovered.

But in the case of DDT, it is very easy to show the good that it has done. It has been estimated, for example, that DDT has saved some 50 million lives since it was introduced; and its effect on raising the quality of human life has been even more important.

DDT kills mosquitoes, and mosquitoes spread malaria. Wipe out the mosquitoes, and you have wiped out malaria. It's as simple as that.

At one time, the World Health Organization (WHO) had hopes of wiping out malaria over the entire globe. These hopes have been dashed, but there are still vast areas of the world where malaria was once a scourge that are essentially free of the disease.

Few Americans are aware of the effects malaria has had on a large part of the world's population. The disease can be fatal, but more often it racks its victim with chills and fevers at intervals of two to four days, and leaves him listless in between. In most of the world, DDT has now wiped out malaria. It has indeed been a great boon to mankind.

On the other hand, much of the DDT used in the U.S. is sprayed on the cotton crop. It is impossible to make a case for using an admittedly injurious chemical to grow a crop that must be bought up by the federal government because it has no real market.

When DDT is used for malaria control, its advantages far outweigh its faults. When it is used for growing unneeded cotton, its faults predominate. No doubt there is a problem. But can one really call it a technological problem? Or even a problem *caused* by technology?

Feeding the World

One of the advances in science that has been in the headlines for the past few years is the development of more-productive strains of grains—corn, wheat and rice in particular.

In 1968, the Philippines was able to produce enough

rice for its population for the first time since 1903! In 1968, Ceylon's rice crop was 13% above the highest in its history. Pakistan's wheat crop beat the previous record by 30%.

For the first time in recent history, people in many lands will have enough to eat. Wonderful, isn't it?

Well, no—not according to Prof. Edmundo Flores, writing in the May-June 1969 issue of *FAO Review* (published by the Food and Agriculture Organization of the United Nations). He claims that this technological breakthrough will lead to rural unemployment and eventual breakdown of the social system in many of the poorer countries.

Flores' point is that the richer farmer, using both the new crops and mechanization, will become productive enough to drive the poorer farmer out of business. And in the countries he is talking about, Flores points out, "He who does not work or has no land does not eat."

This possibility was also noted in an article in the *New York Times* for Apr. 6, 1970. In it, Dr. Robert F. Chandler, director of the International Rice Institute, where high-yield rice was developed, is quoted as saying, ". . . how can it be wrong to increase the amount of food for people who eat and to increase the incomes of farmers?"

Who is right? The next ten years should tell.

Technology Does Good, Too

With all the talk about the damage that technology does to the environment and the quality of life, there is little appreciation of the advantages that it has brought.

I remember listening to a television interview with a student radical who explained that he planned to spend his life in an attempt to overthrow "the system." He was violent in his objections to "technology," which he said was destroying the lives of the people. When asked if he ever intended to take a job, he replied, confidently, "No, I don't expect ever to starve."

What struck me was that this is probably the first time in history that a person could reaonably make that statement. The combination of fertilizers, tractors, high-yield crops, food-processing and transportation technology *will* probably keep him from starving. But obviously this never occurred to him.

Consider the unpleasant work that has been eliminated around the home by the combination of oil-burner, washing machine, and vacuum cleaner.

The flush toilet, which we take for granted, means that nobody has to empty a chamber pot! And piped-water systems mean that no one has to carry water from a well or stream. Efficient transport means that we can eat fresh produce even in the wintertime.

The Phosphate Flap

Take the problem of phosphates in the Great Lakes. They act as a fertilizer that promotes the overgrowth of algae, to the detriment of other plants and animals.

Where does the phosphate come from? Almost all can be traced to three main sources: human wastes (sewage), agricultural fertilizer runoff (rainwater leaching of fertilized soils), and detergents (which use phosphates as builders). They're all the fault of technology.

The concentration of human wastes comes from the concentration of people that make industry more efficient. It also is caused by medical technology that has kept people alive and helped to build up the population. And also, the plentiful supply of food adds to the problem (starving people produce less waste). Eliminating the technology will eliminate the problem. The population will be spread out, hungry, and have a high death rate—all of which will decrease phosphate in the sewage.

Eliminating agricultural fertilizers will have a twofold effect; runoff will be reduced, and a reduced food supply will leave people hungry and may even starve some of them. Again, the result will be less phosphate in the lakes.

Eliminating phosphates from detergents will have little effect except to reduce phosphate in the sewage. Soaps do a reasonably good job of cleaning.

In the case of detergents, eliminating the technology may make sense. In the other two cases (sewage and runoff) it becomes absurd. Here, the obvious answer is more technology. not less.

We'd Be Worse Off Without Technology

Yes, technology is a cause of many problems. And because it is almost impossible to predict the results of new technology, it will cause more problems in the future.

The alternative to technology is an immediate reduction in population and a lower standard of living Somehow it doesn't seem much of a choice.

"As Long As It Doesn't Kill Anybody . . ."

The Story of How a Major Discovery Went Wrong

The criterion of safety that has
dominated the DDT controversy is that
the chemical must not be toxic to
man or other "non-target" organisms.
Trouble is, declares environmental
scientist • Charles F. Wurster • it is
the wrong criterion. The fact that
it took two decades to discover the
error (and may well take decades more
to clean up the damage) is something
of a modern object lesson in asking
right questions and getting right
answers.

DDT was approved for use as an insecticide
in this country about 30 years ago. Although
not much was known about it, it was presumed to be a safe chemical and was consequently used without restriction.
                                 Last year
the U. S. Department of Agriculture changed
its mind—or at least so it appeared. USDA
announced that DDT would be "phased
out," and although little has happened so
far, there are strong pressures from scientists and citizens to stop all use.
                                 To many
people that is probably a puzzling set of
circumstances. If so little was known about
DDT in the beginning, how was it allowed
to come into such widespread use? And if
it is really a hazardous chemical, why did
it take so long to find that out?
                           Although a full
answer to either of those questions is complex, it seems to me that there are clear and
unambiguous explanations for at least
some of the things that have happened
since 1945. I think, for example, that the
original decision on DDT is quite defensible, and while the delay of 25 years is not
defensible, it is at least understandable.

INNOVATION, January, 1971, vol. 17, 21-28.

You
may gather that I am not a neutral observer,
though I started out being neutral. I have
been studying the effects of DDT on birds
and other organisms for some time and I
have long been convinced that we must
stop its use completely in the United States
because it is harmful to birds, fish, and
other wildlife, because it is a hazard to human health, and because adequate or superior alternatives are available.
   Since DDT is rapidly becoming a classic
example of a technological decision being
made without adequate knowledge, it is
instructive for us to examine its history to
try to learn how and why the great experiment went wrong.
                           First let's review briefly
where we stand today.
                           There are four charac-
teristics of DDT that make it more trouble-
some than it might otherwise be. First, DDT
and its closely related breakdown products,
DDE and DDD, are very stable materials.
They are not easily decomposed and may
persist in the environment for many years.
In fact, that is one of the virtues of DDT as
an insecticide; it can remain active for

months after a single spraying.

DDT is also mobile — it doesn't stay where we put it, but circulates around the world in the air and the waters.

Third, DDT is soluble in organic materials and virtually insoluble in water. That means that it will not stay in water but will be accumulated by living tissues where, because of its stability and these solubility characteristics, it will remain for some time.

Finally, it is a very active material biologically and is not at all specific to the particular insect pest. It kills all insects (except those that have become resistant to it). The biological activity also extends to essentially all animals. DDT kills birds, fish, reptiles, amphibians, and mammals if they receive a large enough dose. Although DDT can kill a man, at high dosages, its acute toxicity — its ability to kill directly — is quite low and there are very few cases of humans suffering from acute DDT poisoning.

And that is where the difficulty lies, for it is not so much the more spectacular direct lethality that concerns us; it is the subtle, sublethal effects. Some of these are quite capable of eliminating a species without ever killing a single individual. The story of the great decline of several species of birds of prey is both fascinating and distressing, and it offers many lessons for us to heed in our future actions. I should add that the hazards of DDT are still controversial among pesticide manufacturers and some members of the pest control industry, but are generally accepted by impartial scientists.

The first signs of trouble among carnivorous birds began to appear in the 1950's with reports that something was the matter with rate of reproduction in the peregrine falcon and several other hawks. By 1965 these previously stable populations had collapsed; the peregrine had become extinct over large portions of its range, and the bald eagle, osprey, and several others were also in serious trouble.

Adult mortality was not the problem. Instead, a very low rate of reproductive success was the central difficulty, with a high rate of egg breakage in the nest being the most characteristic symptom. A reader of much of the ornithological literature from 1955 to 1965 might well have concluded that pesticides were involved in those declines, but the evidence was purely circumstantial. The really hard data were just not there.

Then an ornithologist in Britain, named Ratcliffe, who had seen numerous eggs broken in the nests of peregrine falcons, suspected that there must be something wrong with the eggs. He visited museums to measure eggshell thicknesses. He found that prior to about 1946 the thickness of the eggshells had remained constant, but since the late 1940s peregrines have been laying eggs with shells about 18% thinner than formerly. The coincidence with the large-scale introduction of DDT seemed suggestive, but was, of course, still only circumstantial.

When Ratcliffe's findings were published in *Nature* in 1967, other scientists who had been studying enzyme induction through the sixties began to suspect that there might be a connection between their laboratory findings and what was going on in the field.

In June 1968 there was a meeting at the University of Rochester that brought together the lab people and the field people, and things started falling into place. Until that time the people in the lab didn't know what was going on in the field, and the people in the field didn't know what was going on in the lab. That meeting brought them together. It was spectacular to see the story unfold.

Since then the details have been worked out. Controlled experiments have established DDT as the cause, and even the mechanism as it operates in these birds has now been so well established that I doubt whether any serious scientists would quarrel with it.

Very briefly, it is this: Being

65

soluble in organic materials, DDT is absorbed by living organisms from the environment in which they live—from the air, land, and water. As one organism eats another the residues are passed up the food chain. With each step, the concentration increases because the DDT is accumulated and retained by the higher organism. Thus DDT residues reach the highest concentrations at the tops of long food chains—in certain hawks, eagles, pelicans, and sea birds, as well as various large fish.

Assuming that the bird doesn't accumulate so much DDT that it becomes lethal (which occasionally *does* occur), then the major effects involve enzymes. DDT increases the breakdown of the female birds' sex hormones by increasing the activity of certain enzymes in the liver. This alters hormone balance and changes reproductive behavior.

Simultaneously, DDT (and DDE, its breakdown product) inhibits the function of the enzyme carbonic anhydrase, which is required for normal eggshell formation. The result: thin-shelled eggs which break prematurely when the bird sits on them, and few if any young birds. Off the coast of southern California the brown pelicans have essentially been laying omelettes, and a nesting colony of 500 pairs raised exactly *one young bird* in 1970.

Our understanding of the relationship between DDT and avian reproduction is now quite conclusive. It's unusual to have such solid evidence where there is such a subtle question of environmental pollution, and where the cause of the problem is so remote from the effect.

In addition to the enzyme effects, DDT has also been shown to be responsible for other serious long-term biological damage. For example, we know that if you feed rats and mice high levels of DDT, there is a statistically significant increase in the amount of cancer shown by the test animals over controls. We also know by the dominant lethal test that DDT causes mutations in test animals.

Now while there are a great many biological differences between rats and people, the genetic material in both is the same—it's DNA. And we know that all people on earth are carrying residues of DDT. We don't know for certain what it does to people, but we know that it has subtle biological effects on all kinds of organisms. If DDT damages the genetic material and causes cancer in rats, it probably does genetic damage and causes cancer in humans as well. But the only real proof is to use people themselves as test animals, and that turns out to be a very unpopular experiment.

Critics of the tests on mice point out that the dosage levels are typically tens or hundreds of times higher than the environment. But it is impractical if not impossible to run the tests any other way. For statistical significance at the dosage levels found in the environment, the tests would require millions of mice; you would need a building the size of the Pentagon with thousands of employees to test for the carcinogenicity (cancer-causing quality) of one chemical.

The important point is that increasing the dosage does *not* turn noncarcinogens into carcinogens, as has frequently been claimed. There are no examples of a chemical or anything else—aspirin, salt, tomato juice—which given in large doses produces cancer or mutations, and given in small dosages does not. I know of no scientific support for the concept of a threshold above which you get cancer and below which you do not.

So that is where we stand with DDT. It is hardly the panacea it was originally thought to be. It kills various forms of wildlife, reduces their reproductive capacity, interferes with various enzyme systems, and has been shown to be both mutagenic and carcinogenic.

Yet, because it was not directly lethal to man, it was considered "safe." But then, of course, it is not usually lethal to peregrines and pelicans either.

As a result of these findings and unrelenting pressure

from scientists, citizens, and various environmental groups, DDT has become a very controversial chemical. One side attacks it and demands a cessation of its use, while pesticide manufacturers, the U.S. Department of Agriculture (USDA), and the pest control industry generally defend it. USDA has taken several token steps at restricting it, but these have been largely ineffective. They have "cancelled" various uses, but cancellation (which does not mean "ban") merely initiates an endless series of administrative appeals. The appeal procedure does nothing to benefit "the environment." Accordingly, in spite of all the headlines about "banning" DDT, there are almost no federal restrictions on its use.

Several states, however, have taken meaningful action to limit the use of DDT. In Wisconsin, where the Environmental Defense Fund brought scientists from all over the country and abroad to testify in hearings held in Madison, the use of DDT is illegal and the "ban" seems airtight. Rigorous restrictions also exist in California, Arizona, New York, and several other states.

These measures, plus DDT's frequent failure at its intended insect control function, have brought about a decline in usage in the U.S. from about 70 million pounds annually to less than 40 million pounds.

To understand how we got so deeply into this, we have to go back to the beginning, or what is very nearly the beginning, to the first widespread use of DDT during World War II. DDT, a relatively simple molecule, had been synthesized in the 1870s but its insecticidal properties were not discovered until 1939 by a Swiss scientist. So World War II was its first real test.

It was an enormous and almost immediate success. There was malaria in Asia and the Pacific, and typhus in Greece and Italy. DDT just knocked them right out. It saved thousands of lives that otherwise would have been lost to insect-borne diseases.

Here was some-thing that had dramatic properties by any past standard. We didn't know very much about the chemical, but its use was justified and understandable under circumstances where thousands of lives were at stake, to say nothing of a war. It would be asking too much for somebody to have come along at that time and say, "Now look, you'd better check first to see if this is going to be bad for birds or fish. What is it going to do to chromosomes and carbonic anhydrase?" Obviously that's absurd.

So we won the war and DDT played an important role in winning it. It was a miracle chemical, no matter how you looked at it. It had a fabulously good press going for it. Chemical companies sought to make it, and large numbers of entomologists thought the war with the insects was over and that they could do anything with DDT. In 1945 it had momentum—economic momentum and emotional momentum—and who was going to come along then to raise a lot of questions about long-range environmental effects?

There was an explosive growth in the use of DDT in those first few years after the war. Production went from 33 million pounds in 1945 to 106 million pounds in 1951. Everybody used it for everything.

I don't mean to imply that DDT was not tested. Tests were made by USDA, by the manufacturers, and by other government agencies, but most of the tests were primitive. They tested for lethality. They tested for short-term effects. If test organisms survived, then all was considered well. They figured they could pour on DDT an inch thick. Its especially low acute toxicity to man gave a particular sense of security.

The phrase "as long as it doesn't kill anybody ..." is, I think, one of the keys to the story of what went wrong.

The dominant tests for the safety of a chemical were—and frequently still are, for that matter—based largely on mortality. Such a testing procedure doesn't account for long-term biological effects. It doesn't account for what has happened to the peregrine and to other species where the reproductive capacity of an entire species is reduced to a small fraction of normal.

It happens without killing one of them.

With DDT we've got very subtle, sneaky little things going on which can totally demolish a species. These things are not so dramatic as the spectacular kill that has birds dropping out of the trees onto the front lawn—though that has happened. Or piles of dead fish several feet deep, though again, that has occurred.

It is not as though there were no indications of potential trouble in the late 1940s, however. The stability of DDT and the other chlorinated hydrocarbons was well known. So were the solubility properties. There were even published tests in 1947 made by the Food and Drug Administration showing that DDT is a carcinogen. Its broad-spectrum biological activity was well known. Scientists knew that it was not specific to a particular kind of insect, or even to insects in general. It didn't take a great deal of brains to see that a persistent chemical that could kill a frog, a fish, a bird, an insect, or a mouse was eventually going to give trouble with things for which it was not intended.

There was no other chemical in use in 1950 that had all these properties and that was being used in such large amounts as DDT. It was then, it seems to me, that somebody should have been blowing some whistles and saying, "Look, this is fine during wartime, but let's be careful with this stuff. It's going to stay out there a long time and may eventually wind up affecting things other than insects."

But no one did, and that was only one of many misjudgments that permitted the use of DDT to continue.

In 1947 the "Federal Insecticide, Fungicide, and Rodenticide Act" (FIFRA) became law, which if you read it seems to be a fairly decent piece of legislation. It provides some protection for people against the reckless use of chemicals, and it also offers some protection for environmental values and nontarget organisms. But it puts the regulation and registration of these chemicals in the hands of the U.S. Department of Argiculture.

That was fine when pesticides were primarily the problem of agriculture. But with the advent of persistent organic pesticides that have effects far beyond the farm, the problem becomes too broad for USDA. USDA's primary constituency is agriculture and the agricultural chemical industry. It is relatively unresponsive to concerns about fish or birds. Or even human health, for that matter.

Of course, the Food and Drug Administration has jurisdiction to fix tolerance levels in human foods. FDA can put pressure on USDA, but the regulatory process remains entirely with USDA. Similarly, the role of the Department of Interior, or HEW, is purely advisory. These departments are permitted to advise USDA, but USDA makes all the decisions, frequently over the objections of the other agencies.

It must have been back in those early years of FIFRA that tenuous relationships began to develop between USDA and the pesticide manufacturers. After all, if there is a federal agency regulating the registration of pesticides, the pesticide manufacturer is naturally going to become very interested in the policies of the regulatory agency.

Again, it's a subtle thing. Open dishonesty does not seem to be involved. Nevertheless, close personal relationships begin to build, and slight emotional conflicts of interest begin to form—all with the result that the regulatory agency becomes increasingly responsive to the industry it is supposed to be regulating. This is not unique to USDA, but it does happen.

You can see how it arises. We have an agency with a group of perfectly honest people in it being approached socially, professionally, and in other ways by people from industry who are also perfectly honest and respectable. Their industry is at stake. It is being regulated by this agency and it is in their best interest to see that the regulation doesn't hurt them.

Thus we have a climate being created for the sale of a product by the best talent that money can buy—lawyers, lobbyists, public relations men, just plain, friendly, glad-handing fellows—a good supply of literature, and a great willingness to cooperate in every way possible.

Out of this comes an exchange of personnel. Somebody leaving USDA to go to industry is very likely to go where he has connections. And vice versa. Interconnecting webs soon develop. In some cases people may even work for both at the same time. A congressional study last year turned up a man who in one capacity was doing pesticide research paid for by Shell Chemical Company, and in another capacity was also a consultant to Shell, and in still a third role was simultaneously a consultant to the Pesticides Registration Division of USDA. He helped maintain the registration of Shell pesticides over the objections of other agencies.

The central point here is that on one side of USDA—the side that favors the sale of a product—we have heavy pressure and penetration by that industry to see that it is well represented. On the other side, who do we have? Who in the 1940s, the 1950s—in the 1960s for that matter—was representing the fish? Or birds? Who represents human health when HEW has no jurisdiction in the matter? Who, besides a few little old ladies in tennis sneakers and a few bird watchers, is representing the environment in general?

I think now, in 1970, things are finally beginning to turn our way ever so slightly. But it has taken a long time.

One of the first occasions in which the public became aware of the fact that some insecticides have undesirable properties was the absurd fire ant program conducted by USDA in the mid-1950s. The fire ant is an imported insect that is found in the southern part of the U.S. It has been there for about 50 years, and in some regions it is a nuisance.

Beginning about 1956 or 1957, there came a massive, multimillion dollar insecticide campaign in which as much as five pounds of dieldrin ·or heptachlor (chlorinated hydrocarbons, like DDT) were applied per acre by airplane over vast areas of the South. The program was a disaster. It killed almost everything that moved, even including some large animals like sheep, cats, and cattle. Great numbers of birds, reptiles, and amphibians were killed. The dosage was reduced to one pound per acre before they finally gave up altogether.

Except among the people who were there to experience it, the fire ant disaster was not very widely known. It didn't get into scientific journals, but was published by the Department of Interior in various reports and bulletins that don't get wide circulation.

Then in 1962 *Silent Spring* was published. Rachel Carson had read all those reports and when they were cited in her book a lot of people got pretty irritated. She also speculated about long-term effects of DDT and the other chlorinated hydro carbons. She had an impressive grasp of the literature, and had dug out obscure things from all kinds of places.

Silent Spring probably marked the beginning of the public campaign against the irresponsible use of pesticides, a campaign that has been growing steadily ever since. Today such organizations as the science-oriented Environmental Defense Fund (of which I am a trustee) are employing litigation as a means for achieving a regulatory structure more responsive to the public interest. There is a growing constituency that is willing to go to bat for environmental values. So perhaps things are improving, but the going is still mighty slow.

Even now, with a large body of public opinion on the side of environmental values, there are habits to be broken, policies and values that must be changed, new concepts and approaches that must be adapted. There is also a massive job of environmental restoration to be done.

When you understand what DDT and some other insecticides do in a farmer's field you realize that using such chemicals can be a little like getting hooked on a drug.

In a field, a cottonfield, say, there may not be a great diversity of plants but there will still be a considerable diversity of insects, perhaps several hundred species. Of that group maybe half of them are phytophagous insects—that is, they eat plant material. The others are entomophagous, meaning that they eat other insects.

The entomophagous insects are obviously on our side because they help control the phytophagous insects, a few of which are potential crop-eating pests. In the absence of disturbance, the two groups of insects generally live in a state of biological equilibrium. We see, then, that the natural biological diversity is desirable for the farmer because it reduces the likelihood that any one phytophagous insect will be able to erupt to pest proportions.

Often the balance is good enough that there aren't any pest problems at all. All these insects fly around and the whole business churns away, so that no steps need be taken to control pests.

Now and then, out of that group of hundreds of species of insects, one or two get out of control. Often it is a chronic trouble-making species like the boll weevil. Its population density rises above a certain threshold where economic damage is threatened and control measures become necessary. It is important that the farmer understand, however, that the application of control measures when insect populations are below the economic threshold is a waste of his money.

But let's say a broad-spectrum poison like DDT is applied. It kills almost all insects—99% of everything. Now we have an essentially sterile field—a field full of plants but almost no insects.

When the effects of the poison wear off, the phytophagous insects often come roaring back because the entomophagous insects are not there to exert any pressure on their populations. That field, stripped of its insects, has abundant food (the crop) for phytophagous insects and no food for the entomophagous insects. So the one comes back and the other doesn't . . . for a time. The time lag can be disastrous for the crop.

The process of indiscriminate destruction should be called "pest creation" rather than "pest control." Not only do we get a resurgence of the original pest species, but species previously under biological control are elevated to pest status. Where we may have had pest A, we now have pests A, B, C, and D. And on top of that, severe mortality from the insecticide exerts heavy selection for those insects that have traits with survival value when exposed to the insecticide. The ones that survive therefore produce offspring that are resistant to the poison. We soon have tougher pests and more of them than before.

The poor farmer finds himself with headaches like he has never known. The salesman knows nothing better than to recommend that he spray again, probably with a heavier dose. The second treatment generates the apparent need for a third and the third for a fourth. Because of the immense reproductive potential of insects, pest resurgence can occur within a week or two. Often the farmer signs a contract with the insecticide people to spray his place weekly—by the calendar—whether it needs it or not.

Here we have a perfect arrangement (perfect for the manufacturer) where a product generates the apparent need for more of itself. Once the farmer starts, he's stuck. From one-tenth of a pound once a year, soon he's up to one pound 10 times a year. That's 100 times as much insecticide. The farmer is using more and enjoying it less. It's expensive, it decreases his profits, and it's ineffective. To say nothing of what it does to the environment.

The farmer's usual source of pest control information is part of the problem. Instead of seeking the advice of trained entomologists of the state agricultural extension service, farmers all too often go to untrained pesticide salesmen with their insect pest problems. Usually the salesman tells him to use more insecticide: "I guess one pound per acre wasn't enough; you'd better use two pounds per acre this time." Insects and insecticide manufacturers are the ultimate winners while farmers and the environment are losers.

Other factors are also at work. There have been alternatives to DDT for almost 20 years. By 1955 there were many chemicals that didn't have the undesirable properties of DDT and certain other chlorinated hydrocarbons. A good example is methoxychlor. It is a molecule that looks very much like DDT but has just enough of a change to make it unstable. It therefore doesn't have the environmental problems built into it.

But methoxychlor didn't have the press going for it that DDT had. It didn't have the name. Its lack of persistence was also against it because farmers wanted to be able to spray once (at least so they thought) and not have to spray again for a while. And finally, it was more costly. Mass production of methoxychlor would have driven the price down, but probably never as low as DDT. Mass production had already driven the price of DDT so low that it was difficult for anything to compete.

It is still true that no insecticide can compete with DDT on a cost-per-pound basis. For that reason, it is probably necessary that DDT continue to be used in poor countries that can't afford anything else. Where it is a choice between DDT and nothing to prevent malaria in undeveloped countries, then the use of DDT seems appropriate.

Actually, the cheapness of DDT is illusory. Its price per pound is the lowest, but the value of an insecticide is measured in insect control per dollar expended. On that scale DDT often does poorly. But if the rest of its price—its environmental damage—is figured into the equation, then DDT comes out being one of the world's most expensive chemicals.

Despite all the propaganda to the contrary—all the little scare stories about how much the cost of food will go up, or how we will be overrun with pests, starvation, and disease without DDT—I don't believe there is any place in the U.S. where the complete elimination of DDT would create serious problems or cause serious economic dislocations. To the contrary, a sane pest-control policy would benefit agriculture and consumers alike, not to mention the environment. Some food prices might even decline.

In addition to the many alternative chemical insecticides and such new approaches to pest problems as juvenile hormones, sex attractants, infertile matings, and so forth, we have new techniques in which there is an effort to depend less upon chemicals and more upon the biological processes that are naturally at work in the field.

Let's go back to that field in which we had several hundred species of insects more or less in biological equilibrium. Suppose that one of the insects goes above our economic threshold and becomes a pest. Instead of knocking the field flat with a dose of DDT or some other potent chemical, suppose we do something very gentle: We spray a little kerosene on the field, or maybe one-tenth of a pound per acre of malathion or dibrom. Perhaps we kill 70 or 80% of the pests, along with some of the entomophagous insects.

We are still left with a reservoir of entomophagous insects, but the survivors of the pest species are too few to cause economic damage, yet sufficient to serve as a food supply for the entomophagous insects. This gentle treatment does not lead to pest resurgence and may actually reestablish a biological equilibrium that will carry through the rest of the season.

This

approach is known as integrated control. Natural biological controls are integrated with chemical control so that we get the most out of each. Planting and harvesting practices, timing, entomophagous insect releases, and other techniques are ecologically blended into an effective and economical insect management system that maintains potential pest species below their economic threshold. Integrated control is not a wild dream of the distant future, but is already in use in many places in the country.

To use such control, however, agriculture has to establish some new habits, and that is difficult when it has been traveling the pure chemical road for 25 years. To begin with, the farmer has to hire a man who can make an expert examination of his fields, a man who will usually call himself an "applied entomologist." After sampling the insect population in the field, the entomologist tells the farmer what and when to spray, and how much. He may suggest the release of entomophagous insects grown for that purpose in an insectary. Or he may say, "Your field looks fine this week. Just leave it alone."

We need more of these people. Unfortunately, many of those that we have are working either for pesticide companies or for agencies that are partially funded, one way or another, by insecticide manufacturers.

A classic example of pure biological control—where control was established without the use of insecticide—is the story of the cottony-cushion scale. It's a tiny, fluffy-looking insect that destroys citrus crops. It appeared in California about 1868, and by 1887 it had exploded to threaten the citrus industry with total ruin. It didn't respond to the pesticides then available (nor does it now); it has a scale on top if it and is hard to kill.

Because the insect was a native of Australia, USDA sent a man there to find entomophagous insects that keep it from being a pest there. He sent back a number of such insects, one of which was the Vedalia, or lady-bird beetle. The Vedalia beetle, a predator of the scale, turned out to be a winner. Within a few weeks of its introduction onto an infested tree it had cleared the tree of scale. Soon the whole orchard was cleared, and word spread all over California. People came in droves with little cages and cookie jars to get a few beetles for their own orchards. Within a year, the cottony-cushion scale ceased to be a pest in all of California.

But the Vedalia beetle did not eliminate the cottony-cushion scale. It merely suppressed it to a low level, and subsequently, the Vedalia population also declined. It was a case of perfect biological control, where a rare insect is held in check by a rare natural enemy. The crisis passed and everybody forgot about both of them.

Then came DDT in 1945.

There were problems with thrips in the citrus orchards. Farmers sprayed DDT to kill the thrips, but before long they were faced with a horrendous explosion of cottony-cushion scale. DDT had killed the Vedalia beetles, and there was a major crisis, worse than the thrips. Vedalia beetles were so reduced that a market developed for them and they sold for as much as $2 apiece.

DDT was stopped and biological control was reestablished. Within a year the Vedalia had cleared up the scale problem again.

Integrated control—or pure biological control as with cottony-cushion scale—is a thing of beauty. It is cheap and effective, and it avoids environmental problems entirely. There is something intellectually satisfying about understanding the insect and its ecology well enough to be able to manipulate its numbers without upsetting the whole system. The answer usually represents a permanent solution to the problem, in contrast to the pure insecticide technique, which is inherently temporary at best.

It is essential to realize that pesticide manufacturers are in business to sell pesticides, not necessarily to control pests. The two objectives are by no means identical, and may be in opposition to each other. The pesticide application that temporarily alleviates the pest problem while creating the need for future applications is very good business, but the permanent solution to a pest problem that eliminates future markets for the pesticide is very bad business indeed. Industry has no incentive to seek permanent solutions to pest problems, solutions that use less insecticide or maybe even no insecticide. No industry can be expected to reduce its own markets intentionally. It is a little like the nylon stocking that doesn't run or the Chevrolet that goes 500,000 miles. What do you sell next year? In some ways our whole sociological structure is out of whack with biological reality. We should be putting our brains and talent into finding Vedalia beetles. That's what USDA should be doing. They do some of that, but not enough. A good bit of what they're doing amounts to simply protecting pesticide manufacturers.

Again, we are beginning to see some hopeful signs. Pesticide registration has been taken away from USDA and put into the new Environmental Protection Agency. We will just have to see how that will work. What concerns me is that the government is taking the same people with their rigid pro-pesticide perspectives and moving them under another roof. Of course, just having them under another roof might change things.

Other changes should be made. The whole structure of advisory committees and consultants to government agencies should be modified. Those people who administer grant funds should be buffered from the source of the funds to protect their objectivity. Some grant funds come inevitably from industry, but we must stop this arrangement where the pesticide industry makes grants to universities for pesticide research, thereby sapping the intellectual resources that should be going into integrated control.

That's what is happening now. Through the channeling and flow of funds, we've got a network of departments of entomology and agriculture, along with state and local agricultural people all over the country, hooked together by a financial web that leads through USDA to pesticide manufacturers. Many of these groups get their grant funds directly from the pesticide companies, so it is understandable that they would reflect the pesticide companies' point of view.

I am not suggesting that we throw out this whole structure, but only that it is desperately in need of some major surgery if we are going to keep capitalistic instincts from upsetting the ecosystems on which we all depend.

The lessons of DDT have implications that extend far beyond pesticides. One of the most obvious is that biological systems, for all their seeming adaptability, ruggedness, and resistance, are still rather delicate things. We cannot broadcast biologically active, persistent chemicals widely into the environment and not expect to have effects. Just because a substance doesn't kill people is no reason to conclude that it is harmless. We have to be more aware of subtle, sublethal effects.

I wonder, for example, what the increasing level of noise is doing to people, especially city people. We know what it does to our hearing: It deafens us in slight but measurable ways. But what kind of psychological damage does it do? What does it do to our decision-making processes? Our general well-being? Noise may very well be another one of those subtle, sneaky things. It won't kill us but it could have serious consequences. Yet few people are even asking questions about noise, let alone doing anything about it. Similar things are happening all around us. A frightening aspect of these ecological issues is that often nobody understands the long-term effects of our actions. We don't know what happens to a civilization that breathes in particles of lead (from automobile exhausts) for 50 or 100 years. We know that mercury can be harmful to biological systems, but we don't really understand what it does, how it works, or how much it takes to have long-term effects.

We have to develop a suspicious nature. We can't assume automatically that everything is fine, that we can fly an SST and it isn't going to hurt anything. We know it will make sonic booms, but we don't know what the booms will do to people, to birds' eggs, to whale communication, to buildings, or anything else. It will put small particles and water vapor into the upper atmosphere, we don't begin to understand that problem. All of us will have to live with the SST and we have a right to ask serious questions and demand adequate answers.

I wonder if we are asking the right questions about these things in 1970 any more than we asked the right questions about DDT in 1945.

IN DEFENSE OF DDT AND OTHER PESTICIDES

by *Norman E. Borlaug*

Nobel Peace Prize 1970

ONCE again the "Naked Ape"—homo sapiens stands at a crossroad. Before deciding along which road to proceed, he hesitates and glances behind at the long road he has trod.

He is both amazed and proud of the tremendous progress he has achieved as he manœuvred and advanced along the pitfall-laden trail of human survival during the brief period he has inhabited the planet Earth. Within the last second—representing a short 5 million years as measured on the geologic clock—he remembers emerging from the bush somewhere in south-east Africa, standing upright on his back legs and beginning to assume the role of Desmond Morris's so-called "Naked Ape".

With a club in one hand and a rock in the other, he stalked animals and became a carnivore. For a long time he struggled for survival as a hunter and food gatherer under the hostile environmental pressures dispensed by a fickle mother nature. More than once he barely averted extinction. He remembers having seen certain other species perish, because of their inability to adjust to the capriciousness of the environment. They have left for posterity only a fragmented history of their existence recorded in the book of fossil rocks.

During his long early period as a hunter, social progress was negligible. Survival itself was man's only sense of achievement. Then his helpmate, neolithic woman, only about 9,000 years ago, invented agriculture and animal husbandry. This brought new hope. It ensured his food supply and lightened his load. Survival became less of a problem. The quality of human life improved. It gave him time for pleasure and time to think. There was time to develop his intellect, a society and, subsequently, a culture.

He made many worthwhile scientific discoveries that made life more comfortable, gratifying and enjoyable. Among these were the discoveries of control for many diseases that had plagued and scourged him from the beginning of time.

With the control of diseases and the resultant drop in death rates, combined with a reliable food supply, human population growth soared. The population monster looms ahead and now threatens to obstruct further progress, and even to erode progress already achieved, unless tamed.

All these events that have taken place along the tortuous road of human progress over which the Naked Ape has travelled, flash back to him now, as he hesitates and reflects be-

Reproduced from THE UNESCO COURIER, February 1972, pp. 4-12.

fore turning forward again to make the necessary decision as to which road to choose, at this, the most complex series of crossroads and intersections that he has ever encountered along the highway of social evolution.

He realizes, as he now meditates his decision, that the world civilization has split into two factions—the privileged world of the developed nations and the improverished world of the developing nations. One is living a life of luxury never before experienced by man outside the proverbial Garden of Eden; the other still leads a life of misery built on poverty. Discordant voices cry out to counsel him now concerning his choice at the next crossroad.

One of the siren songs is that of the so-called establishment which counsels continuing on the road that has brought the high standard of living to the developed or privileged nations. This infers that these benefits can be extended to those in the under-developed world by following the same path.

A second tone cries out against the establishment and the materialistic world, even while they themselves are living as social drop-outs, social parasites in reality.

The third, and most seductive voice, is that of those extreme environmentalists who discredit science and advocate a back to nature movement. They demand the discontinuation of the use of chemical compounds—even though they are absolutely essential for protecting man against diseases, and for restoring fertility to the worn out soil so man can produce his food, and protect his crops against the ravages of weeds, diseases and insects.

Civilization as it is known today could not have evolved, nor can it survive, without an adequate food supply. Yet, food is something that is taken for granted by most world leaders, despite the fact that more than half of the population of the world is hungry and an even larger proportion malnourished.

With the help of our science we must not only increase our food supplies, but also ensure them against biological and physical catastrophes, through international efforts involving both developed and developing nations.

International granaries for food reserves financed by all nations should be established for use in case of need. These granaries should be strategically located in different geographical areas so as to simplify logistics in time of emergencies. And these food reserves must be made available to all who need them—and before famine strikes, not afterwards.

Man can and must prevent the tragedy of famine in the future instead of merely trying with pious regret to salvage the human wreckage of the famine, as he has so often done in the past. We will be guilty of criminal negligence, without extenuation, if we permit future famines. Humanity cannot any longer tolerate that guilt.

The destiny of world civilization depends upon providing a decent standard of living for all mankind. It has been said that "Universal and lasting peace can be established only if it is based upon social justice. If you desire peace, cultivate justice". Almost certainly, however, the first essential component of social justice is adequate food for all mankind.

I feel that the aforementioned guiding principle must be modified to read: "If you desire peace, cultivate justice, but at the same time cultivate the fields efficiently to produce more bread; otherwise there will be no peace."

During the past five years, spectacular progress has been made in increasing wheat, rice and maize production in several of the most populous developing countries of southern Asia, where widespread famine appeared inevitable only six years ago. Most of the increase in production has resulted from increased yields of grain per hectare, a particularly important development because there is little possibility of expanding the cultivated area in the densely populated areas of Asia.

The term "Green Revolution" has been used by the popular press to describe the spectacular increase in cereal grain production during the past five years. Perhaps the term Green Revolution, as commonly used, is premature, too optimistic, or too broad in scope. Too often it seems to convey the impression of a general revolution in yields per hectare and in total production of all crops throughout vast areas comprising many countries.

These implications both oversimplify and distort the facts. The only crops which have been appreciably affected up to the present time are wheat, rice and maize. Yields of other important cereals, such as sorghums, millets and barley, have been only slightly affected; nor has there been any appreciable increase in yield or production of the pulse or legume grain crops, which are essential in the diets of cereal consuming populations. Moreover, it must be emphasized that thus far the great increase in production has been in irrigated areas.

Nevertheless, the number of farmers, small as well as large, who are adopting the new seeds and new technology is increasing very rapidly, and the increase in numbers during the past four years has been phenomenal. Cereal production in the rainfed areas still remains relatively unaffected by the impact of the Green Revolution, but significant change has been made in several countries during the past three years.

Despite these qualifications, however, tremendous progress has been made in increasing cereal production in India, Pakistan and the Philippines during the past three years. Other countries that are beginning to show significant increases in production include Afghanistan, Algeria, Brazil, Ceylon, Indonesia, Israel, Iran, Kenya, Malaysia, Morocco, Thailand, Tunisia and Turkey.

Before attempting to evaluate the significance of the Green Revolution one must establish the point of view of the appraiser. The Green Revolution has entirely different meanings to most people in the affluent nations of the "privileged world" than to those in the developing nations of the "forgotten world".

Most people in industrialized societies have difficulty in comprehending and appreciating the vital significance of providing high-yielding strains of wheat, rice, maize, sorghum and millet for the people of the developing nations.

There are no miracles in agricultural production. Nor is there such a thing as a miracle variety of wheat, rice or maize which can serve as an elixir to cure all ills of a stagnant, traditional agriculture. Nevertheless, it is the Mexican dwarf wheat varieties and their newer derivatives that have been the principal catalyst which triggered off the Green Revolution.

If the high-yielding dwarf wheat and rice varieties were the catalysts that ignited the Green Revolution, then chemical fertilizer was the fuel that powered its forward thrust. The responsiveness of the high-yielding varieties has greatly increased fertilizer consumption. The new varieties not only respond to much heavier dosages of fertilizer than the old ones but are also much more efficient in its use.

The continued success of the Green Revolution will hinge, however, upon whether agriculture will be permitted to use the inputs—agricultural chemicals —including chemical fertilizers and pesticides, both absolutely necessary to cope with hunger. If agriculture is denied their use because of unwise legislation that is now being promoted by a powerful lobby group of hysterical environmentalists—who are provoking fear by predicting doom for the world through chemical poisoning—then the world will be doomed but not by chemical poisoning, but from starvation.

The current vicious, hysterical propaganda campaign against the use of agricultural chemicals, being promoted today by fear-provoking, irresponsible environmentalists, had its genesis in the best-selling "half-science, half-fiction novel" *Silent Spring*, published in 1962.

This poignant, powerful book—written by the talented scientist Rachel Carson—sowed the seeds for the propaganda whirlwind and the press, radio and television circuses that are being sponsored in the name of conservation today.

It is both sad and unfortunate that *Silent Spring* was the last book which was written by this gentle, great scientist and authoress. She had previously contributed so much to the understanding of the beauties of nature in the best-sellers *Under the Wind* and *The Sea Around Us*.

The gravest defect of *Silent Spring* was that it presented a very incomplete, inaccurate and oversimplified picture of the needs of the interrelated, worldwide, complex problems of health, food, fibre, wildlife, recreation and human population. It made no mention of the importance of chemicals such as fertilizers and pesticides for producing and protecting our food and fibre crops. Nor did it mention that by producing more food per unit of cultivated area more land would be available for other uses, including recreation and wildlife.

Certainly the greatest inexcusable error of omission was that of neglecting to mention the valuable role DDT has played in bringing malaria under control in many countries.

Silent Spring convinced the general public that the use of pesticides—and especially DDT—was upsetting the "balance of nature" and was doing great damage to wildlife, especially birds and fishes. It implied that a number of species were facing extinction because of its use.

Moreover, it left the impression that agriculture really did not need insecticides if it changed its methods. It indicated that farmers by adopting a system of extensive mono-culture, have made their crops more vulnerable to pests than necessary.

According to this expert, farmers have compounded their errors more by applying insecticides in attempting to kill insect pests and in the process have generally only killed off the predators, parasites and pathogens that normally kept the insect under control, and thereby only further upset the "balance of nature."

Moreover, according to the author, insects have invariably soon developed resistance to the insecticide. It implied that, by shifting to other suitable insect control measures already available, the losses from insects could be kept under control without chemicals.

I am in complete agreement that we should try to preserve all forms of wildlife as part of our heritage, as far as it is possible to do so. On the other hand, let us not become egotistical to the point of assuming supernatural powers. A glance at the book of rocks tells us of the impotency of many species, including man against the forces of nature. Yet it is incredible that only a few, if any, of the leaders of the current environmentalist movement have studied palaeontology and the "parade of the species," in the geologic past.

Spencer estimates that 99 per cent of all the species that have lived, since the candle of life was first lit on the planet earth about 3.2 billion years ago, have flunked the adaptation imperative: "evolve or perish," and consequently have now become extinct.

The implied command: "evolve or perish" has been an unwritten natural law from the beginning of time. It is equally evident in the physical and biological world.

The multitude of changes in the physical features of the earth, as well as in our solar system itself, have repeatedly greatly modified the environment of the earth. Climates have changed time and again in many parts of our world. Vast areas that once possessed tropical climates have subsequently been covered by continental ice sheets. Areas that once were blessed with heavy rainfall have become desert and vice versa. These changes in environments have, in turn, exerted strong selection pressure on the evolution of all forms of life.

There are undoubtedly many subtle changes being exerted on the environment of the planet today that are beyond the influence and control of man. Man too, however, is exerting strong influence on the environment. The composite effect of the present day selection pressure of the environment, affected both by natural and human influences, will undoubtedly continue to take its toll of some species that are poorly adapted to the current world environment.

Rhodes, Zim and Shaffer estimate that there are at present approximately 1,100,000 species of animals, many of them very simple forms, and 350,000 species of plants that currently inhabit the planet earth. Of these, the United States Fish and Wildlife Service in 1966 listed 33 species of mammals, 49 species of birds and 9 species of reptiles and amphibians, and 38 species of fish in the U.S.A., which were either rare or endangered.

In discussing the causes for reduction in numbers and possible disappearance of these 129 species, the destruction of the habitat and disturbances resulting from man's activities were paramount. Pesticides were mentioned as possible contributing factors in only two cases. In the past three or four years there has been much propaganda, but little convincing scientific evidence, put forward by environmentalists indicating that DDT has contributed to the decline of the Bald Eagle, Peregrine Falcon, American Osprey and Californian Condor.

One does not need a thin egg shell hypothesis due to DDT to explain the reduction in the population of these species. The truth of the matter is that many ornithologists had reported on the reduction in populations of these large birds of prey as far back as the 1880s and 1890s, long before the time of DDT. It is almost a foregone conclusion, for anyone who uses some common sense, that one or more of these species is about to flunk the imperative "evolve or perish." Their habitats are being destroyed by the encroachment of man.

Protective legislation alone will not, in most cases, be adequate to save them. Dynamic research, propagation and good sound game management might do so, providing human population pressures on their habitat are not too great.

Although it is generally the long-term continuing changes in the environment which exert their effects on the evolution and survival or the extinction of a species, there are many other changes in the environment that effect the more short-term "balance of nature," among the many species in a given habitat or ecosystem. These are the seasonal shifts we are concerned with in producing and protecting our crops or animals.

The cliché "In balance with nature," which is in common usage today by modern environmentalists is very misleading. It implies we would have a favourable "in balance with nature" to assure the protection of our crop species if the "balance of nature" were not upset by man. This, of course, is not true. Nor is there in existence a single "in balance with nature" ecosystem. Rather there is, within a given area, an infinite number of local and many more extensive merging ecosystems.

None of them are in static equilibrium. They are in a constant state of dynamic change, responding to the changes in the environment. At different times, the selection pressure provoking change is drought, floods, frosts, heat, insect or disease attacks, or invasion of the habitat by other species.

EARLY in my career as a forester working in a large primitive or wilderness area completely isolated from the influence of man, I learned of the fickleness of nature. I have seen 20 forest fires ignited by a single "dry thunder (electric) storm". Some of these fires started by lightning destroyed or damaged vast areas of several forest types (ecosystems).

In the same area I have seen tens of thousands of acres of lodgepole pine killed by Dendroctonus spp., infestation. The havoc done by the Dendroctonus beetle should not have happened according to some pseudo-ecologists, for it was, after all, a native insect pest with its entire army of natural predators, parasites and pathogens, and consequently should have been "in balance with nature."

Many times I have seen attempts made to grow cotton without the use of insecticides in the native home of the boll weevil in Mexico where all of the native predators of this insect were present. The results were disastrous.

In fact, it was difficult to tell from casual observation whether the cotton was being grown for the production of fibre to clothe man or for the production of feed for a native insect. Nevertheless, there should have been, according to our environmentalist jargon, an "in balance with nature" equilibrium.

I must also point out that modern agriculture—with 3,700 million people demanding food and fibre—has no choice but to grow extensive areas to a single crop in areas ecologically best suited to the culture of that crop. This was not true 5,000 years ago, when there was less population pressure, so that crops could be grown in small isolated fields.

It therefore becomes abundantly clear that we cannot rely on biological control alone to protect our food and fibre crops from the fickleness of nature. If we leave things to mother nature's whims, we will harvest only one third or one half of the yield per unit of cultivated area that can be harvested using a modern balanced technological package of practices.

Dr. Knipling has clearly indicated that we must, for the forseeable future, continue to use an integrated approach to control the insect pests of man, of the crops and of the animals on which he depends. Insect control is a complex problem for there are more than 200 insects that are or have been important on our main crops, animals and forests. We will need to use an integrated approach to hold them in abeyance.

It is true that in the past few decades spectacular control of a few insect species have been obtained with biological, bio-environmental or other non-chemical methods i.e. cottony-cushion scale of citrus, the spotted alfalfa aphid and the screw-worm of livestock in Florida.

Some day it may be possible to use alternate non-chemical methods to control many of the insects responsible for the most severe crop and animal losses, but that day, if ever attainable, lies far in the future.

Today, however, conventional insecticides are needed to control 80 to 90 per cent of the insect problems affecting agriculture and public health. Meanwhile, research to find new techniques and methods, must be strengthened. Present control programmes must be designed to take advantage of the best materials and techniques available so as to reduce losses to an acceptable level.

The environmentalists would now like to have a legislative ban placed on DDT so as to prohibit it for any use in the U.S.A. Almost certainly as soon as this is achieved, these organizations will begin a worldwide propaganda barrage to have it banned everywhere. This must not be permitted to happen, until an even more effective and safer insecticide is available, for no chemical has ever done as much as DDT to improve the health, economic and social benefits of the people of the developing nations.

The World Health Organization (WHO), with the assistance of the Pan American Health Organization and the United Nations Children's Fund (UNICEF), in 1955 launched a worldwide campaign against malaria, based on spraying the interior of all houses with DDT, so as to kill the Anopheline vector and break the cycle.

Of the 124 countries and territories in the tropics where malaria has existed, the disease has been eradicated from 19. There are 48 other countries in which eradication programmes are in progress and an additional 37 where extensive control programmes are under way. There remain only 20 nations in malarial areas where no programmes have yet been initiated.

THERE is also dramatic evidence from Ceylon of what can happen if a programme is stopped before eradication is accomplished. When the campaign was initiated in the mid 1950s there were more than two million cases of malaria in Ceylon. By 1962 it had dropped to 31 cases and by 1963 to 17, at which point the spray programme was discontinued for budgetary reasons. By 1967 the number of cases had jumped to 3,000 and by 1968 to more than 16,000. Before the programme could be re-established, in late 1969, two million cases had reappeared.

In summarizing the progress in this world wide malaria campaign on February 2, 1971, officials of WHO made the following statement:

"More than 1,000 million people have been freed from the risk of malaria in the past 25 years, mostly thanks to DDT. This is an achievement unparalleled in the annals of public health. But even today 329 million people are being protected from malaria through DDT spraying operations for malaria

control or total eradication.

"The improvement in health resulting from malaria campaigns has broken the vicious circle of poverty and disease resulting in ample economic benefits: increased production of rice (and wheat), because the labour force is able to work, and the opening up of vast areas for agricultural production.

"The safety record of DDT to man is truly remarkable. At the height of its production 400,000 tons a year were used for agriculture, forestry, public health, etc. Yet in spite of prolonged exposure by hundreds of millions of people, and the heavy occupational exposure of considerable numbers, the only confirmed cases of injury have been the result of massive accidental or suicidal swallowing of DDT. There is no evidence in man that DDT is causing cancer or genetic change."

Although more than 1,400 chemicals have been tested by WHO for use in malarial campaigns, only two have shown promise and both of these are far inferior to DDT.

As more and more scientific evidence accumulates, the charges against DDT become less and less convincing. There is evidence, of course, that man and most species of birds, fish and animals that have been examined have small quantities of DDT and/or other related componds such as polychlorinated bi-phenyls in their fat. But there is very little convincing evidence available to date which indicates that it is threatening the existence of any species, nor is it causing any discernible injury to man.

Part of the past confusion concerning pesticides in the environment derives from the tremendous improvements that have been made in recent years in chemical analysis. With gas chromotography it became possible to detect 1 or 2 parts per billion, or even a few parts per trillion, both of which, of course, would have gone unnoticed 20 years ago. But such sensitive methods can also detect contaminants and in the hands of inexpert operators may lead to wrong conclusions.

A recent article by Dr. Thomas H. Jukes, a reputable biochemist, emphasizes this dilemma: "How reliable is the test? There has been a great hue and cry over alleged traces of DDT in the Antarctic penguins, amounts of the order of 1 or 2 parts per billion. I have not yet been convinced of the validity of these results."

A few months ago at the University of Wisconsin, some soil samples that had been sealed since 1910 were tested for synthetic organochlorine pesticides by the latest, most delicate gas chromotographic procedure. Several pesticides were detected in 32 of the 34 samples. The only flaw was that these pesticides not only were not used in 1910, they did not even exist until 1940.

Another complication is that the residues of a class of modern compound called polychlorinated bi-phenyls (P.C.B.'s) interfere with the DDT test. The P.C.B.'s are used in water-proofing compounds, asphalt, waxes, synthetic adhesives, hydraulic fluids, electrical apparatus and general plastics. They are widely distributed in the fat of wildlife species, in which they have originated as industrial wastes taken up by aquatic species. To sum up, P.C.B.'s are not used as pesticides, but they interfere with pesticides residue analysis and they are toxic.

Another complicating factor in identifying the origin of chlorinated hydrocarbons in human, animal, bird or fish tissue is that many thousands of tons of chemical wastes of all kinds have been and are still being dumped into the oceans. Do not some of these also get into the food chain, even if they still have not got into the "hysterical word chain"?

IT is now obvious that the current aim is to ban DDT, first in the USA, and then in the world if possible. But DDT is only the first of the dominoes. But it is the toughest of all to knock out because of its excellent known contributions and safety record.

As soon as DDT is successfully banned, there will be a push for the banning of all chlorinated hydrocarbons, then in order, the organic phosphates and carbamate insecticides. Once the task is finished on insecticides, they will attack the weed-killers, and eventually the fungicides. As a matter of fact, by default, they have already been successful in having organic mercury seed disinfectants and slimeicides for papermills banned.

This ban was achieved during all of the confusion resulting from finding mercury in fish, first in fresh water species, in the Great Lakes, and rivers of the USA, and subsequently in both tuna and swordfish. The ridiculousness of some of this rhetoric came to light recently when someone analysed tuna caught 90 years ago and found it contained about the same level of mercury as those caught today.

Moreover, it has been shown that swordfish recently caught in ocean waters hundreds of miles from possible industrial contamination contained 1 to 2 p.p.m. of mercury. This indicates clearly that both tuna and swordfish are picking up the low levels of mercury from the ocean food chains, of which this metal has always been a part.

If the use of pesticides in the USA were to be completely banned, crop losses would probably soar to 50 per cent, and food prices would increase 4 to 5 fold. Who then would provide for the food needs of the low income groups? Certainly not the privileged environmentalists.

Within the past decade, because of the improved technology and higher yields, it has been possible to remove 50 million acres from cultivation and still meet both the domestic and export needs for agricultural products. Were the USA still relying on the 1940 technology, however, not only would the 50 million acres now held in reserve be back under the plough, but an additional area of 241.9 million acres by necessity would have been opened to cultivation. In reality it would require considerably more than 241.9 million acres of additional land since the quality of the land would be poorer than that now in cultivation.

In order to bring under cultivation an additional 241.9 million acres (and perhaps considerably more because of the poorer land quality) it would be necessary to open to cultivation lands that were in a large part rolling or semi-arid, and consequently vulnerable to erosion by water and wind.

It would also require clearing the forests from large areas so as to meet the food, feed, oil and fibre needs of the nation. Now reflect on the additional havoc that this expansion of cultivated area would do to wildlife habitat, and especially on rare and endangered species of animals and birds that are already on the brink of extinction.

Looking at it from another angle 291.9 million acres of land, an area roughly equivalent to the total land area of the USA east of the Mississippi river and south of the Ohio river, is today available for other uses, because of the improvements in crop production technology that have taken place in the past 30 years. These uses include recreation, wildlife and forestry.

It behoves all mankind to increase the efficiency of agriculture throughout the world if we wish to alleviate human suffering, conserve wildlife and improve recreational opportunities. Unless the food production of East Africa is expanded to meet the growing food needs, the large animals in the Game Reserves of East Africa will be poached out of existence within the next three decades. Similarly, the elephant, tiger, and peacock will perish from India because of population pressure.

It is hard to understand why the conservation organizations and environmentalists have taken a negative rather than positive view in trying to protect wildlife. Why have they not promoted research and fought for more funds for game management in general? Why have they not fought for more funds for research so that qualified scientists can be assigned to study the reasons why certain threatened species are on the verge of extinction, and whether it is feasible to try to save them?

Why do they not spend more of their energies and funds on educating the public on the adverse effects of population pressure and rampant population growth on wildlife and the environment? How many of the US public, for example, know that more than 100,000 deer are killed each year by automobiles, whereas everyone is informed in the press or television whenever a few birds or fish are found dead, presumably—but this is not necessarily proven—from a pesticide?

Imagine the rhetoric that would be produced if 100,000 deer were killed by a pesticide.

I have been a great admirer of the spendid work that has been done by game management experts in the USA in re-establishing species, such as the wild turkey, that were nearly extinct. Under wise management and protection many other species of wildlife have made spectacular comebacks. The tremendous success of the introduction of the Chinese ring-neck pheasant, the Hungarian and Chukkar partridge, are other tremendous accomplishments. The research that has brought under control the lamprey, that threatened the survival of the lake trout is another tremendous achievement; so is the introduction of Coho salmon into Lake Michigan and Lake Superior. The breeding of faster growing and more sturdy salmon is another tremendous step forward.

I repeat what I have said many times before: without thinking, conservationists and environmentalists and only partially-informed people in the communications media have embarked on a crusade designed to end the use of agricultural chemicals, such as pesticides and fertilizers. They give no thought to the end result of such action: the eventual starvation and political chaos that will plague the world.

HERESY IN THE HINTERLAND--1966

D. Neslo

The origin and history of measurement is a fascinating study.
In ancient times standards of measurement were closely associated with the human body, or closely allied with some human
activity. As might be expected, the more archaic means of
measurement were usually the least accurate.

The "cup of tea" measurement of Tibet is a case in point.
This unit of measurement may be loosely defined as the distance
a Tibetan peasant could run with a steaming cup of tea before
it cooled to a drinkable temperature. As you can well imagine,
all sorts of variables detracted from the accuracy of this means
of measurement.

We are more concerned, of course, with the origin of our own
means of measurement. One or two examples will suffice to illustrate the initial inaccuracy of this system.

For instance, the English yard was decreed by King Henry I
in 1101 as the distance from the end of his nose to the tip of
his second finger when his arm was extended horizontally. The
English rod and foot, as described in a 16th century German
treatise, were ascertained as follows:

> Stand at the door of a church on Sunday and bid 16 men
> to stop, tall ones and small ones, as they happen to leave
> at the end of the service. Make them put their left feet,
> one behind the other, and the length thus obtained shall be
> the right and lawful rood [rod] to measure and survey the
> land with, and the 16th part of this measure shall be a
> right and lawful foot.

Considerable refinement has been brought about since the
"foot" unit of measurement was first determined. Interestingly
enough, the refinement is based upon internationally established
metric units, which were standardized nearly one hundred years
ago.

In Paris, on May 13, 1874, a bar of 250 kilograms of platinum iridium metal was cast. From this alloy, prototypes of the
meter were to be made but tests revealed existence of impurities. Another bar of alloy with fewer impurities was cast in
England (home of the pound, foot, and year), and from this cast-
ing 30 metric bars were drawn. Each bar's precise length was
based upon a unit which was one ten-millionth part of the distance between the north pole and the equator. These international metric standards are now stored with infinite care in
the International Bureau of Weights and Measures near Paris.

JOURNAL OF INDUSTRIAL ARTS EDUCATION, 1966, Vol. 26, pp. 37-39.

Even this precision was not sufficient, and on October 14, 1960, scientists adopted as a new unit the wavelength in vacuum of the orange light given off by the gas Krypton 86 when a current is passed through it. This new standard is consistently accurate to one thousandth-millionth of a meter.

. . . .

It is inevitable that the metric system will some day be adopted in America. Adoption is long overdue. A universal form of measurement is one more major factor which may be used in promoting mutual understanding and effective communication among literate people everywhere. In America, we have shackled ourselves with lowered efficiency, unnecessary inconvenience and limited sales in clinging to an ancient system of measurement.

With the recent conversion of England to the metric system, almost every nation with the exception of America is riding the metric tide. Thus England, the very nation from whom we adopted our system of measurement, has now converted to a more workable standard. Developing nations of the world such as India and Egypt have chosen the metric system. Although conversion from one system to another may be a prolonged procedure, India, with a minimum of industrialization, was able to convert within a period of five years. As long as almost all other nations are already on the metric system, it behooves our nation to adopt and take advantage of a simplified form of measurement and, concomitantly, get in step with world usage.

. . . .

It has been a hundred years since Congress adopted metric dimensions as the system of measurement. Adoption took place in 1866 and since that time the metric system has been used almost exclusively in science laboratories and other areas of our economy. It is primarily in industry where we have adopted the English system with such blindness and dedication. There are a host of practical reasons why our nation should, in the very near future, change all measurement to the metric system:

1. It is commonly used in trade transactions in most countries except our own. Thus, communication on the international level is subsequently limited by interpretation.

. . . .

2. The metric system can readily be applied to manufacturing processes with resultant simplicity of procedure and uniformity. Unwieldly, simple fractions such as 3/16, 5/16, 11/32, and 7/64, not to mention unwieldly combinations such as improper fractions, become a thing of the past. Decimals are much simpler and, subsequently, more accurate to work with. The most staunch supporters of the English system have to admit that metric measurements reduce the time and amount of calculation

and immeasurably cut costs in our modern and complex technology.

. . . .

3. The metric system is an interrelated technique that is applicable to linear, mass, and standards of volume. It is because of this interrelatedness that students may easily learn to understand this system early in their childhood.

4. It is universally used in scientific studies and actually facilitates such research. Without its universal applicability many international studies could not be made. Its inherent capacity for infinite extension into the atomic or space worlds makes it essential for our further pursuance in order to learn more about nuclear energy and space exploration.

The metric system is already dominant in American science. For example, almost all of the drug industry has already adopted this system and, in fact, has used it for many years.

5. Since the meter is approximately an arm's length or a long step, we need not feel that we are divesting the existing association between the human body and lineal measurement.

6. Deception in buying and selling would be greatly reduced if the entire world were on the metric system.

Even though the commonly-used English name weights and measures of the United States are a distinct system, they are legally defined only in terms of the agreed-upon international metric standards. The obvious, objective question to ask then, is, "If our system is defined on the basis of international metric standards, why not simplify the whole process and use the metric system?"

. . . .

LEAD, THE INEXCUSABLE POLLUTANT

by **PAUL P. CRAIG**

If the crust of planet earth were to be chopped into a million pieces, somewhere between ten and fifteen of them would consist of the chemical element lead. As far as scientists have been able to discover up to now, lead contributes nothing to the development or maintenance of life, either in plants, or in animals, or in man. On the contrary, the evolutionary process that brought forth the human species seems to have recognized long ago that lead is poisonous to life; the farther upward one searches in the chain of species that feed upon other species, the less lead is found.

Yet, within the degree of accuracy to which such matters have been measured, the scientific indications are that the surface waters of earth's oceans today contain ten times as much lead as they did before the human animal emerged.

And the American people today are carrying around in their bodies one hundred times the amount of lead they would have absorbed from a primitive environment.

What does this mean?

Simply that man has changed his natural environment to such an extent and has employed lead in making the changes in such a way as to systematically poison himself.

SATURDAY REVIEW, October 2, 1971, 68-70 ff.

Originally, all the lead on earth was buried in the planetary crust. Man began digging out the metal about 5,000 years ago, probably after finding it accidentally in the ores from which he obtained silver. Tin also was present in silver-bearing rock, and could be mixed with lead to form pewter and so provide a protective coating for copper pots and pans that otherwise poisoned the food prepared in them. Lead was likewise popular with potters, who used it as glazing for ceramic vessels.

The poisonous effects of lead on the human organism have long been recognized. The early Romans, in their quest for silver, smelted large amounts of ore that contained lead. About 400 tons of lead were recovered for each ton of silver. The mining and smelting were performed by slaves, who undoubtedly often died of lead poisoning.

The lead was used for a wide variety of purposes, including roof sheathing, and cooking and wine vessels. Democritus noted that the acidity of wine could be reduced by the addition of lead oxide. Pliny specified that leaden pots must be used in making grape syrup; dissolved lead apparently improved the flavor of the syrup.

Since the ruling classes had most access to leaden vessels, they constituted the group that was most poisoned. The resulting decline in their birth rate and in their creative and governing ability has been documented impressively by Dr. S. C. Gilfillan in an ingenious piece of detective work [see *SR*, Aug. 7, 1965].

Centuries after the Roman Empire collapsed, apparently without understanding what had happened to it, pioneers of modern American civilization acted to prevent a repetition of the performance. Governors of the Massachusetts Bay Colony in New England outlawed the distillation of rum in leaded vessels in order to prevent what were then called "the dry gripes." Generations of boys on both sides of the Atlantic played with toy lead soldiers until the toys were shown to be connected with sickness and the death of children who nibbled on them.

During the early years of the twentieth century, lead poisoning was common among house painters. The most characteristic symptom was wrist drop, a tendency of the wrist muscles to sag. The sagging resulted from lead interference with the nerves that control the muscles.

The early mortality of painters and workers in the lead processing industry was relatively high and easily identified. On diagnosis, action could be taken to eliminate the sources of exposure. The most decisive action was banning lead from interior paint some years ago. This protected the painters, but not the infants and toddling children who picked off and ate paint peeling from neglected walls. New coats of lead-free paints blocked off the danger residing in the old leaded paints, but when the new coats wore thin and were not in their turn covered, the underlying lead paints again came within reach of the children's hands.

Convulsions, delirium, coma, severe and irreversible brain damage, blindness, paralysis, mental retardation, and death can result from lead poisoning. In children, the early symptoms are particularly subtle. Victims become irritable, sleepy, or cranky. They may be troubled either with diarrhea or its opposite, constipation. Only if a pediatrician is looking for lead poisoning is he likely to identify it in a child, for most children are frequently irritable, sleepy, or cranky. So, although lead poisoning is known to be one of the major sources of injury to young children in low-income families (two hundred die every year in America alone), lead's impact on other children can only be surmised.

Most lead that enters the human body does so through food. This lead, which enters the stomach, is rather inefficiently absorbed by the body, and only about 5 to 10 per cent of the lead ingested actually enters the blood stream. Inhaled lead is far more serious, for the fine particles emitted by automobiles are retained within the alveoli of the lungs and are absorbed by the body with an efficiency of about 40 per cent. Thus, a small quantity of lead inhaled can do far more damage than a large quantity consumed. By emitting lead into the atmosphere, man has bypassed complex and effective mechanisms designed by nature to keep the lead burden of humans low.

Because of these circumstances, the emission of lead through the exhaust pipes of internal combustion-engined automobiles has become man's greatest worry in connection with lead poisoning—greatest because it has been growing constantly since 1923, when lead was first introduced as an additive to automotive fuel. Although concentrated in the cities and hence visited most heavily on city dwellers, atmospheric lead is carried by the winds and deposited all over the globe. The index of its presence is the lead content of the Greenland icecap, which has been traced back to 800 B.C. and shown to have been explosively accelerated during the last half century.

The total daily intake of lead in the food and drink of an individual American is typically about 300 micrograms, of which 15 to 30 micrograms is absorbed. The average city dweller experiences an atmospheric lead level of about 2 micrograms per cubic meter. He inhales about 20 cubic meters of air per day, of which 40 per cent, or 16 micrograms, is absorbed. Thus, at least one-third of the total lead absorbed by average American urban dwellers arises directly from atmospheric lead.

In unfavorably situated cities, the concentration of atmospheric lead can be substantially higher than the levels just mentioned. In midtown Manhattan, for example, average values of 7.5 micrograms of lead per cubic meter of air have been reported. Lead content of some city dust approaches 1 per cent, which is equal to the proportion of lead found in some ores. Grass harvested from alongside highways has been found to contain as much as one hundred times the lead concentration of grass not exposed to automobile exhaust. Recently, at the Staten Island Zoo, two leopards were paralyzed, a horned owl's feathers dropped out, and a number of captive snakes lost their ability to slither. All proved to be victims of lead poisoning, and the source of the lead was the grass, leaves, and soil in outdoor cages, as well as the paint on the cage bars. Dr. T. J. Chow of Scripps Oceanographic Institute recently reported that in San Diego average values of lead are now 8 micrograms per cubic meter of air; he noted that the concentrations are rising at a rate of 5 per cent per year. There can no longer be any question that atmospheric lead is at a dangerous level.

The degree of the danger cannot be stated precisely without an accurate measure of the amount of the lead burden now being carried by the bodies of Americans and of the margin between this level and that known to produce crippling or fatal effects. If the margin is small, it is important to search for subtle effects that would not be noticed in a conventional public health survey.

The most commonly used indicator of exposure to lead is the concentration of lead in the blood. There is at present a narrow margin between the average blood lead level in Americans and the level associated with severe poisoning. The level considered diagnostic of lead poisoning in healthy males is 0.8 parts of lead per million (ppm) parts of blood. Today the average American's blood lead concentration is about 0.2 ppm—one-fourth of the amount commonly considered hazardous to adults and almost half the level indicative of acute poisoning in children.

At best, the margin of safety concept is questionable; at worst, it can be disastrous. The definition of the margin depends in large degree upon the sophistication of diagnosis. Safety levels are set so that known deleterious effects do not occur—at least not often. As diagnostic techniques improve, effects in individuals can be detected at lower levels. As statistical techniques improve, it becomes possible to search for subtle effects in large populations, as well as for synergistic effects in which the sensitivity of the body to a particular insult is increased due to the presence of some other pollutant, dietary deficiency, or the like. With increased sophistication, one also can detect groups of people who are especially sensitive. In the case of lead, it is essential that the most sensitive group —the children—be given particular emphasis in setting permissible criteria and standards.

There has developed in recent years a considerable body of data indicating that a margin of safety for lead exposure may not exist at all and that damage may occur even at low exposure levels. If this is the case, it is imperative that unnecessary exposure to all types of lead be held to an absolute minimum.

Experiments with animals offer an excellent approach to the search for low-level effects. In an elegant series of tests carried out over many years in a special low-lead-level laboratory, Dr. Henry Schroeder of Dartmouth College has found that chromium-deficient mice carrying lead burdens typical of those found in the American people have reduced life spans and increased susceptibility to disease. Chromium deficiency is thought to occur in many humans. Recent experiments in Russia have shown that rabbits exposed to atmospheric lead at levels not much different from those found in some U. S. cities exhibit various functional disabilities and pathological anomalies.

Detailed statistical studies are necessary to delineate the extent of these subtle effects, which may consist of a diminishing of intelligence by a few points, a decrease in nervous coordination and mechanical dexterity, or a general rundown feeling.

In contrast to many other pollutants, lead is a cumulative poison. Studies of Americans show that the older a person is, the more lead is concentrated in his body. (A slight decrease occurs in persons over sixty years of age.) The total body burden of lead in middle-aged Americans is about 200 milligrams, of which about 90 per cent is concentrated in the skeletal structure.

As the impact of lead effluent upon our health and our economy becomes recognized, the need for controls is increasingly evident. The most stringent of these will have to be adopted by the United States, which now consumes about 1.3 million of the total world lead consumption of 2.2-million tons. Some of this consumption—the part that goes into electric batteries, solder, and pewter—can be recycled at the end of the useful lifetimes of those products, but lead used as a gasoline additive cannot be recovered. It can only be prevented from entering the atmosphere in the first place.

The impact of atmospheric lead emitted from automobile exhausts constitutes a threat to health so severe that on this count alone lead emissions should be prohibited. However, the major pressures for the elimination of lead from gasoline so far have not resulted primarily from this important concern, but rather from the fact that lead in gasoline interferes with the control of many other automobile emissions.

Of the many methods proposed to control the amount of hydrocarbons, carbon monoxide, and oxides of nitrogen leaving the exhaust pipes of automobiles, one of the most discussed is the catalytic converter. The catalytic converter depends on the filtering capability of porous material with a large surface area in proportion to the volume occupied. The pores in the material fill rapidly with lead particles when leaded gasoline is used, and the process of converting the other pollutants to their harmless constituents is blocked.

To end the blockage, major auto makers equipped most of their 1971 models with low-compression engines able to operate on 91-octane lead-free gasoline. This surprise action forced the oil companies to shift petroleum refining methods. Several of them have introduced low-lead and lead-free gasoline throughout the country.

Low-compression engines get fewer high-speed miles per gallon of fuel. Also, there is some evidence that emission of aromatic hydrocarbons increases as lead content of fuel falls. These points have been argued forcefully by the Ethyl Corporation and by Du Pont, the primary makers of tetraethyl lead, the gasoline additive. Pollution control devices that will operate on leaded gasoline can be built, they say. As alternatives to catalytic converters, they have demonstrated prototype thermo-reactors, which are claimed not to foul when leaded gasoline is used. Theoretically (but not yet practically), lead particulates can be removed from the exhaust stream by special filters and separation devices. Continued use of lead in gasoline, Ethyl and Du Pont contend, will provide needed engine lubrication and avoid a controversial phenomenon called "valve seat pound-in," which may cause rapid wear of valve seats in cars using unleaded gas.

From an environmental point of view, the Ethyl and Du Pont approaches to the lead problem cannot be rejected or ignored. The environmentalist is primarily concerned with what comes out of the exhaust pipe rather than what goes into the gasoline. Reaction processes occurring in the internal combustion engine are complex and poorly understood. What is essential is that automotive emissions be controlled as expeditiously as possible, using the best technology currently available.

While the gasoline suppliers are now providing low-lead gasoline, it costs several cents per gallon more than leaded gasoline of the same octane rating. Cost-conscious motorists therefore are avoiding the new fuel. They will have to be encouraged to buy it. The encouragement could come through governmental regulation. With official standards in force, all manufacturers would compete on an equal footing. Unfortunately, such standards have yet to be set, although the Environmental Protection Agency has promised them for mid-December. By that time, a year and a half will have passed since the Environmental Defense Fund's petition to the U.S. Department of Health, Education, and Welfare for establishment of criteria and standards for lead.

President Nixon has demonstrated his concern over lead poisoning by ordering all federally owned vehicles to operate on unleaded gasoline; however, because of bulk buying under long-term contract, the order cannot take practical effect until the next fiscal year. The President has also proposed a tax on lead, but Congress has not been enthusiastic about enacting one.

Spurred by the 1970 Clean Air Act amendments sponsored by Senator Edmund S. Muskie of Maine (they authorized consideration of environmental health effects of fuel additives), the Environmental Protection Agency has issued several reports on the danger of lead pollution and on the economics of its removal. One recent document, published in August, said lead-free gasoline could be made available across the country by 1975 at an additional cost to the motorist of between 0.2 and 0.9 cents per gallon of fuel.

The EPA published in June an interim report on a massive study of lead in several major cities. A comparison between atmospheric lead levels measured in 1961-62 and again in 1968-69 showed that ambient levels had increased by 13 to 33 per cent in Cincinnati, by 33 to 64 per cent in Los Angeles, and by 2 to 36 per cent in Philadelphia.

The results of this important and alarming EPA study were unfortunately omitted from a National Academy of Sciences' study of lead released early in September, which concluded from older information that "the lead content of the air over most major cities has not changed over the last fifteen years." The omission of the most recent and best study from the NAS report led the Academy to recommend in its final conclusions only that more research is required on the health impact of automotive lead. Subsequently, the press reported that the Academy study showed that "lead is an overstated peril" (*The Wall Street Journal*, September 7). Meanwhile, the President's Council on Environmental Quality reported that there is little doubt that, at the present rate of pollution, diseases due to lead toxicity will emerge within a few years. The National Academy lead study is a dramatic example of how our most prestigious scientific body is incapable of taking a stand regarding the risks associated with introduction into the environment of substances that damage people in insidious epidemiological ways.

It seems fair to say that if we did not now have lead in our gasoline, and if some gasoline manufacturer proposed to add lead to his product, his proposal would be denied. We are clearly risking our health for the sake of cheap speed on the road. Although lead in gasoline does not constitute the most severe threat to health and well-being confronting society, it is one of the most unnecessary threats—hence an unforgiveable one; yet, one we must all bear.

AMERICA'S ANSWER TO THE POPULATION CRISIS

Fred A. Olsen

A top-level government study of 1972 predicted a population
of 271 million by the year 2000 predicated on an average 2.1
children per woman. If each woman were to bear an average 3.1
children during her reproductive years, the population of the
United States would number some 322 million by the turn of the
century. In any event, the spectre of overpopulation is not
pleasant to behold.

Unless the population growth of the nation is held to a minimum, a whole host of related problems must be faced and resolved.
There may be:

1. Overwhelming demands made upon our recreational resources.
2. Fewer jobs and/or lower per capita income.
3. Continued overcrowding and congestion of our metropolitan
    and suburban areas.
4. Increased pressure upon availability of food supplies.
5. Greatly increased depletion rate of all natural resources,
    particularly minerals and water.
6. Untold demands made upon our sources of energy.

A 1971 census indicated that women between the ages of 18-24
expected to bear an average of 2.4 children during their lifetime. Unless those anticipations of 1971 have changed significantly, the United States may likely have a population in the
year 2000 approaching 300 million.

Liberalized abortion laws, news media announcements, education
and religion may all effectively assist in minimizing population
growth. Japan, for instance, has employed these techniques to
stabilize their population. America, however, has another technique of birth control, somewhat unique to this nation, discrim-
inatory only by age, shocking, casually accepted by the general
population and, therefore, most opprobrious. Enter--the automobile.

The brief statistics of accidents which follow will serve to
illustrate that a great number of the 50,000-plus people who are
annually killed on our nation's highways are relatively young.

On May 3, 1971, nine young people perished in a car/train
crash near Covington, Georgia. The nine ranged in age from
twelve to sixteen. It was reported that the secondhand automobile was bent nearly in half by the impact of the train, with
the right rear end of the car lacking only five feet of touching
the right front wheel.

On October 24, 1971, eleven young people were killed in a
head-on collision in the eastern part of Washington state. Five
ORIGINAL MANUSCRIPT, 1973.

of those were from seventeen to nineteen years of age.

On the same day in Portland, Oregon, nine young people died in a head-on collision. Six of those were from seventeen to twenty years of age.

On June 7, 1973, in the state of Ohio, six young people died in a truck/automobile collision. All were seventeen years of age.

While isolated examples have been chosen in order to accentuate the incidence of young people who are slaughtered on our highways, the occurrence itself is not uncommon. A search of state patrol records would undoubtedly reveal hundreds of similar accidents and others of greater magnitude occurring over the years.

Thus, it is alarming to note that 54,700 people were killed in 1971 in motor vehicle related accidents. Of those 54,700 fatalities, approximately 23,000 were under 24 years of age. Put another way, about 43% of all automobile related fatalities of 1971 involved persons under 24 years of age. Through our callous acceptance of the automobile as our primary mode of transportation we have developed an effective, but ghastly system of combatting population growth. That is, we rear the young to a procreative age and then either destroy or maim thousands of them annually via the automobile before they or their offspring can become part of the population explosion pattern.

Americans have always been an inventive and innovative people. As a matter of fact, this trait has been one of necessity for they have been coping with the vagaries of nature and deprivation for hundreds of years, and inventiveness has become a way of life. Thus, according to Mishan, our most notorious creation has been the automobile. Americans, on a per capita basis, own more cars, drive farther and more frequently, and kill more people by this means than the people of any other nation. Quite a record!

Whether our marriage to the automobile has been a worthwhile and profitable union is a question that many people are now beginning to ask. We are all aware of the conveniences of this vehicle. Most notable, of course, is the ease and frequency with which we can get about.

Interestingly enough, we have never been greatly concerned with the unbelievable human suffering that has resulted from accidents in which the automobile has been the proximate cause. However, we are gradually becoming aware of the magnitude of damage that the automobile is imposing upon our environment.

Other Facets of the Stone

Over one half of all air pollution is caused by automobile exhaust. One need only stand at a busy signalized city intersection to gasp and grasp the enormity of this pollution problem. If all the automobiles were removed from the American scene today, America's air would be relatively clean and clear. An effort is being made to reduce the amount of pollutants given off in exhaust but it will be years before significant national progress is made.

In deference to the automobile we have set aside vast amounts of acreage in order that the rusting hulks can fade away into oblivion as obnoxiously as possible. Seemingly, there is little that can be done to rid the scene of these carcasses for the steel companies are unable to reprocess the metal for one reason or another. Apparently there are too many contaminants such as rubber, fabric, nonferrous metals, glass, etc. Oddly enough, foreign countries without the iron ore resources that we have are able to import these carcasses from America, salvage the metal, utilize it in manufacturing and export those products to America at a profit.

It would be so easy to correct the problem by paying an additional amount for the automobile at the time of purchase--an amount sufficient to cover the expense of salvage. The additional fee that the buyer of the new car would have to pay for salvage would be negligible.

Several years ago the city of Portland, Maine, undertook the project of locating, removing, and disposing of auto hulks that lay rusting away throughout the city. It was a magnificent community effort! The Bayside Residents Neighborhood Center helped in the location of the auto bodies and obtained the necessary owner releases. In addition, the Portland Jaycees collected the vehicles. Others joined in the community effort where necessary. The cars were crushed and returned to the steel mills and/or exported. It cost the city approximately $3.00 each to remove the 400 hulks that they were able to locate.

As one looks at the cities of America today, one is prone to wonder whether cities exist for the automobile or whether they exist for people. Automobiles have literally destroyed the city, making what was once a cohesive unit into a sprawling megalopolis. Vast amounts of acreage are being consumed by asphalt and concrete. It is estimated that over 50% of the total area of Los Angeles is surfaced for streets, parking lots, or service facilities of one sort or another. In order to accommodate the automobile in what was once the city, huge and ugly monolithic multi-story garages have had to be built. Other valuable land has been set aside for street-level parking facilities. In many cities the curb lane has been set aside for parking. Service

stations and their oil slicks, fumes, servo-mechanisms, and
clamor exist everywhere. Other valuable space has been utilized
for new and used car sales facilities and automobile supply
houses. That we have made such a concerted effort to accommodate the automobile and so little effort to provide comfort for
city dwellers and/or shoppers, is almost incomprehensible.

It is interesting to note, in addition, that roadways are
frequently built along the most scenic property within the city,
such as along rivers, around lake shores, and along salt water
frontage. Thus, man with his myopic vision, has created an
effective barrier between the people and their water wonderlands.
Examples abound, but one need only refer to the Alaskan Way Via-
duct (a raised roadway) and Alaskan Way (surface roadway) in
Seattle, both of which effectively separate the city from the
charm and attraction of Elliott Bay.

The mobility that comes with automobile ownership has made
it possible for the city dweller to do his shopping in suburbia.
Thus, shopping centers have been built about the environs of
all the larger cities. In the process, the city loses some of
its central core and the megalopolis effect takes over. In a
very short time there will be one vast, amorphic shopping center
existing between Washington, D.C. and Boston. Along with this
will be the inevitable traffic jams.

Many foreigners who visit this nation are overwhelmed by the
absurdity of one individual driving about in a multi-hundred
horsepower vehicle. Of course, the foreigner is not aware that
the American motorist thinks of the automobile as considerably
more than a service vehicle. It can transport the motorist
from one destination to another and, of course, this is important. In addition, it is a status symbol, an escape mechanism,
and a means of expressing individuality. Perhaps Americans
have made a fetish of this deadly weapon because it is their
one opportunity in this frustrating life in which they may rule
supreme over "something."

It is estimated that one track of transit can accommodate as
many people as 20 lanes of roadway and virtually eliminate the
need for downtown parking facilities. Our Federal government
spends 100 times more money on highways than on mass transit.
In fact the Federal government is virtually in collusion with
the multiplicity of facets involved in the automotive industry
for the support of that industry. Advocates maintain that the
support is justifiable for the automotive industry contributes
over 10% of the gross national product and employs approximately 12,000,000 people.

While we are unwilling to recognize its demise as inevitable,
the automobile and its mass utilization that <u>we</u> <u>know</u> <u>today</u> must

disappear from the American scene. It is not at all possible
that we can continue to dissect our cities for the purpose of
constructing freeways or destroying our valuable rural property
by paving it. In the construction of superhighways we have allowed other means of transporting people to degenerate while we
have concentrated our entire effort on means of accommodating
the motor car. Neither the automobile manufacturers nor the
highway contractors are wholly responsible for this emphasis.
The American people are to blame for allowing this tragedy to
occur.

Alternatives

As mentioned previously, enormous amounts of Federal and
State monies are expended in highway construction. Little financial support is given to the mass transit systems of our
major cities, most of which could be extensively overhauled in
order to better attract and accommodate more passengers. Mass
transit has been successful in some cities. In Philadelphia,
over two-thirds of the commuters use the transit system.

It is one thing to complain about the evils of the automobile
and quite another to suggest something better in its stead.
Americans have become used to the joys of mobility and are not
likely to retire to the home and fireplace in lieu of being able
to move about often, quickly, and without restriction. However,
there must be a better means of moving people about this nation
without putting overwhelming emphasis upon the motor car.

Mass transit is one means of solving the problem where there
are sufficient numbers of people that the transit system may
operate at a profit. Proponents have deisgned carveyors, transit expressways, and guid-o-matic trains. In addition, under
consideration are tubes through which one could move forth and
back across the country at incredible speeds.

Another proposed means of local and more personalized transit
is that of DART (demand actuated road transport). Such a system operates on neither route nor schedule and responds only to
the demand of potential customers. Its manner of operation
could be somewhat as follows: The customer phones in for service and his request is logged into a computer according to the
number of passengers involved, point of origin, and destination.
The computer selects the most appropriate vehicle (minibus)
available. This selection is determined on the basis and location of the minibuses, the number of passengers on board each,
and their destinations. After this information has been sorted
the computer may automatically notify the customer just prior
to pick-up time. Such a service could cost less than one-third
the cost of a taxi and do away with the necessity of personal
automobile use within the city. It is believed that the DART

97

system could operate with efficiency in both semi-rural and residential areas. Another possibility is that DART need not take the passengers all the way into the city but, instead, deliver them to collection points where mass transit may be used.

Technology may be a great boon to mass transit systems, as is evident in the modern, plush passenger train running between Tokyo and Osaka, Japan. The run is approximately 3 hours during which time this highspeed train covers 300 miles. Every facet of the train's operation is computerized and managed from a control center in Tokyo.

A computer operated low capacity vehicle designed to move people within a city is scheduled to go into service in Morgantown, West Virginia. It is called Personal Rapid Transit (PRT) by the Department of Transportation. Morgantown is an ideal city for the first practical test of PRT. The main campus is located downtown, and the city is so clogged with traffic congestion that the students are unable to schedule consecutive classes between the downtown and outer campus. Eventually, a fleet of more than 70 electrically powered, rubber wheeled vehicles will roll along the 3.2 miles of ground level and elevated guideways linking the downtown with outer campus buildings.

Stockholm has a coordinated mass transportation system that is relatively inexpensive. For about ten dollars a ticket holder is entitled to travel by the carrier of his choice whether bus, subway, train or any combination anywhere within the greater Stockholm area for one month. For fifteen dollars a month, a ticket holder may include in his coordinated travel plans the use of Stockholm's ferries along with the land carriers.

Montreal's "metro" is quite innovative in several respects. The waiting stations are spotlessly clean, and local businesses have donated pieces of art for the stations. In addition, the cars run on tires that are filled with nitrogen. Thus, the ride provided is similar to that of walking on thickly piled carpet.

Air travel in America has become a time-consuming process. This has developed because of the problems of getting to and from the air terminal. The air travel time is usually quite modest. Much of this frustration and time loss can be alleviated by the utilization of air busses, helicopters, monorails, and other means of mass surface transportation. The problem cannot be solved by building more roadways to the airport, for then the matter is compounded by the inadequacy of parking area.

Other means of mass transportation might be accomplished by the hovercraft. Hovercrafts are already in operation between

England and France.

Another proposal is that of tube transport. A train in such
a tube could be moved at speeds up to and beyond 500 miles per
hour. Such tubes could be located above or preferably below
ground.

Possible alternatives to transportation via automobile are
legion. However, two things are certain. We cannot return to
the "good old days" of horse transport. At the turn of the
century, many of the very problems now heaped upon the automobile were attributed to the horse. There were pollution problems, noxious odors, and an almost unbelievable racket of horses'
hooves and carriage wheels on cobblestone streets. Writers of
the period demanded banishment of the horse because of the
aforementioned problems as well as others, such as disease,
manure disposal, swarms of flies and related insects, and disposal of dead horses.

It is also inevitable that other means of transportation must
be found to augment and supplant the "horseless carriage."

BIBLIOGRAPHY

Mishan, E. J. "On Making the Future Safe for Mankind." The
Public Interest, pp. 33-61, Summer, 1971.

Statistics Division Accident Facts, Chicago: National Safety
Council, 1972.

Tarr, Joel A. "Urban Pollution Many Long Years Ago." American
Heritage, 22:65-69, October 1971.

United States Department of Commerce Population Characteristics,
Series P-20, No. 232, p. 4, February 1972.

Welles, Chris. "Bitterest Fight: New Mass Transit vs. More
Highways." Life, May 12, 1967.

"What the Birth Rate Means to the America of the Year 2000,"
U.S. News and World Report, 72:45-46, March 20, 1972.

What Price Ecology?

KENNETH S. TOLLETT

In the early nineteen-sixties the civil-rights movement propelled itself into the public consciousness by a series of sit-ins, wade-ins, ride-ins, and other forms of dramatic protest and demonstration. The 1954 school desegregation case had earlier reaffirmed the American dream of equal justice and equal opportunity for all; but the conscience and consciousness of the country were not fully aroused until Martin Luther King's non-violent demonstrations forced the entire nation to face anew the outrage of black second-class citizenship and oppression.

Although two civil-rights acts were enacted during the somnolent Eisenhower years, the 1964 Civil Rights Act finally put Congress and the President dramatically upon record as being fully committed to the spirit and substance of the 1954 school desegregation case.

Of the many lessons learned from the civil-rights movement, none has more potential for both good and evil than the realization that one is not listened to in this country unless one shouts; there is no response unless one raises hell. Indeed, affluent complac-ency will deal with issues and problems only when they are escalated into a crisis. We are, whether we like it or not, a crisis-oriented society.

It is understandable, then, that the serious issues and problems involving our environment are labeled as a crisis. How else can the attention of the country be obtained without crying wolf? It is this aspect of human character, especially the Anglo-American mind, that causes me to approach the ecology hysteria with more than a bit of cynicism and skepticism.

The peace movement, learning from the civil-rights movement, escalated its tactics until it obtained speciously dramatic responses: bombing halts, Selective Service reform, and troop withdrawals. It is just a question of time before the ecophiles begin to escalate their tactics in drawing attention to the ecological crisis. In the meantime the race, poverty, urban, and Vietnamese problems have been pushed into the background. Feminism is already in the wings to replace or upstage the ecology fad.

It is obvious that I do not believe the racial crisis should be ignored, and

THE CENTER MAGAZINE, July/August, 1970, vol. 3, 20-21.

that I have serious doubts whether the ecology fad will continue long, or if it does continue long whether it can be followed without ignoring the racial crisis. This does not mean I think the race problem is the most important problem facing mankind. The most important problem facing mankind is survival. And there are a number of issues creating the survival problem, some of which are the arms race, nationalism, ideological fanaticism, and racism. In both the short and long run, the first of these problems threatens human survival more than the other three combined. Thus if we were a rational people, we would devote more energy to dealing with that problem than with any other. Nationalism is very difficult to deal with directly, and ideological fanaticism has the longest unbroken record in history of causing war and mischief to mankind. Racism's history is parallel to ideological fanaticism. Indeed it has reënforced and complemented both excessive nationalism and ideological fanaticism.

Because of the difficulty in dealing with the above four threats to survival, many say the seriousness and universality of the ecology problem may be a means of indirectly dealing with the former four. Thus the ecology crisis is a fifth issue involving the survival of mankind which may be used as a means of circumventing nationalistic chauvinism, ideological fanaticism, and anti-human racism. I hope so.

However, I submit that a society cannot live long with rank injustice without becoming desensitized to the atrocities and outrages of war. If we adjust ourselves to famine, poverty, ignorance, and suffering when they are manifested in our daily lives the threat of their being magnified in times of impending war or ecological crisis does not disturb our complacent indifference to them. The United States of America has adjusted itself this way to the remorseless mistreatment of blacks and other minority groups; it is not especially concerned in a human way with the possible awful consequences of the arms race and ecological disaster.

To paraphrase a statement of Immanuel Kant, as long as there is anywhere in the world widespread deprivation, exploitation, and injustice, no other part of the world is safe from them in one form or another. Racial and social injustices in America have desensitized us to the horrors of war and undermined our power to recognize the universal dignity of mankind. Our shameful unconcern with the aspirations of the deprived masses south of the border and all over the world has placed us in a position of having to wage a counterrevolutionary policy in order to preserve ourselves, as we see it, from the threat of North Vietnamese imperialism. I am afraid that the vestiges of this unconcern can even be found in the rhetoric of eco-talk.

Ecophiles and eco-escapists feel that in order to protect life on the planet, animals and the environment must be elevated while man is deflated. They want to avoid exterminating any animal species in order to preserve the variety and stability of the biosphere, or to preserve gene pools. How do you think aesthetic snobbery of this kind sounds to ghetto residents — black or white — who are told that roaches, rats, and bedbugs must not be exterminated in order to preserve the variety, stability, and fascination of the biosphere? There is no doubt that there is a contradiction in a movement to preserve a viable environment which entails the belittlement of mankind. It is not much different in human sensitivity and sensibility than Congress spending thirty billion dollars to send a man to the moon, but thinking it is extravagant to spend thirty million dollars to control rats in the ghettos.

There are even some rumblings that economic expansion, because of its probable adverse effects upon the environment, may have to be stopped. This of course means that large numbers of blacks and members of other under-classes will be permanently consigned to poverty unless there is a violent revolution and redistribution of wealth in this country. Obviously, the only way a better standard of living can be given to the millions of Americans, black and white, who are ravaged by poverty, disease, and malnutrition is for our economy to continue to grow, perhaps not as fast as before but certainly fast enough to permit those at the bottom to share in the fruits of this economy without unduly taking away from those who already have the good things of our society.

It is difficult for anyone to persuade me that he really has a deep concern for the dignity of man when he equates the need for a general respect for life which entails concern for condors, rare trees, animals, and birds with the need for respect for human life. It takes no thought at all and very little experience to realize that there have been many, many racists and lynchers who kept their gardens well and took even better care of their pets.

Although I regard the environment problem as serious, then, I do not place it high among my priorities. This because I believe if the pace of excitement about the environment and the quality of life continues unabated, there will be a cop-out on the other priorities. A few redwood trees, condors, whooping cranes, buffaloes, and other picturesque fauna and flora may be preserved for the aesthetic satisfaction of cultural snobs and elitest intellectuals; a few signboards may be removed from highways, and industrial development here and there may be thwarted (to the detriment of the workingman

and for the benefit of the leisure class's conspicious concern for scenic beauty, unoiled surfing waters, and clean and silent air for their luxurious estates, homes, and patios); but at the same time hardly anything will be done to clean up the slums, improve medical services for the lower and lower-middle classes, continue expansion of higher educational opportunity, deal with mass transportation problems of our urban centers, and maintain the struggle for economic, racial, and human justice.

In short, I am of the opinion that the environment movement is a fad which permits President Nixon and affluent white America to escape from dealing with the issues involving human survival that I have already enumerated.

Student unrest, the postal strike, and the various teacher strikes tell us that people who have been comparatively neglected are not going to continue to permit temporizing, pussyfooting, procrastination, and evasion in the name of economy, balanced budgets, and inflation. If this is the case for them, then you can imagine how blacks view the ecology fad. It is the greatest insult to them to talk about preserving scenic beauties and keeping bays and oceans clean while Americans won't provide for adequate garbage collection in poor and black communities.

The ghetto environment has been a physical and economic disaster area for years. America has done little about it up to now. There is nothing in the ecology movement to suggest that anything significant is going to be done about it through eco-talk and eco-escapism. What is necessary is not a new animism or pantheism, but an authentic humanism which is unequivocally committed to the sanctity, dignity, and integrity of human life.

The Tragedy of the Commons

Garrett Hardin

At the end of a thoughtful article on the future of nuclear war, Wiesner and York (*1*) concluded that: "Both sides in the arms race are . . . confronted by the dilemma of steadily increasing military power and steadily decreasing national security. *It is our considered professional judgment that this dilemma has no technical solution.* If the great powers continue to look for solutions in the area of science and technology only, the result will be to worsen the situation."

I would like to focus your attention not on the subject of the article (national security in a nuclear world) but on the kind of conclusion they reached, namely that there is no technical solution to the problem. An implicit and almost universal assumption of discussions published in professional and semipopular scientific journals is that the problem under discussion has a technical solution. A technical solution may be defined as one that requires a change only in the techniques of the natural sciences, demanding little or nothing in the way of change in human values or ideas of morality.

In our day (though not in earlier times) technical solutions are always welcome. Because of previous failures in prophecy, it takes courage to assert that a desired technical solution is not possible. Wiesner and York exhibited this courage; publishing in a science journal, they insisted that the solution to the problem was not to be found in the natural sciences. They cautiously qualified their statement with the phrase, "It is our considered professional judgment. . . ." Whether they were right or not is not the concern of the present article. Rather, the concern here is with the important concept of a class of human problems which can be called "no technical solution problems," and, more specifically, with the identification and discussion of one of these.

It is easy to show that the class is not a null class. Recall the game of tick-tack-toe. Consider the problem, "How can I win the game of tick-tack-toe?" It is well known that I cannot, if I assume (in keeping with the conventions of game theory) that my opponent understands the game perfectly. Put another way, there is no "technical solution" to the problem. I can win only by giving a radical meaning to the word "win." I can hit my opponent over the head; or I can drug him; or I can falsify the records. Every way in which I "win" involves, in some sense, an abandonment of the game, as we intuitively understand it. (I can also, of course, openly abandon the game—refuse to play it. This is what most adults do.)

The class of "No technical solution problems" has members. My thesis is that the "population problem," as conventionally conceived, is a member of this class. How it is conventionally conceived needs some comment. It is fair to say that most people who anguish over the population problem are trying to find a way to avoid the evils of overpopulation without relinquishing any of

SCIENCE, Dec. 13, 1968, vol. 162, pp. 1243-1248.

the privileges they now enjoy. They think that farming the seas or developing new strains of wheat will solve the problem—technologically. I try to show here that the solution they seek cannot be found. The population problem cannot be solved in a technical way, any more than can the problem of winning the game of tick-tack-toe.

What Shall We Maximize?

Population, as Malthus said, naturally tends to grow "geometrically," or, as we would now say, exponentially. In a finite world this means that the per capita share of the world's goods must steadily decrease. Is ours a finite world?

A fair defense can be put forward for the view that the world is infinite; or that we do not know that it is not. But, in terms of the practical problems that we must face in the next few generations with the foreseeable technology, it is clear that we will greatly increase human misery if we do not, during the immediate future, assume that the world available to the terrestrial human population is finite. "Space" is no escape (2).

A finite world can support only a finite population; therefore, population growth must eventually equal zero. (The case of perpetual wide fluctuations above and below zero is a trivial variant that need not be discussed.) When this condition is met, what will be the situation of mankind? Specifically, can Bentham's goal of "the greatest good for the greatest number" be realized?

No—for two reasons, each sufficient by itself. The first is a theoretical one. It is not mathematically possible to maximize for two (or more) variables at the same time. This was clearly stated by von Neumann and Morgenstern (3),

but the principle is implicit in the theory of partial differential equations, dating back at least to D'Alembert (1717–1783).

The second reason springs directly from biological facts. To live, any organism must have a source of energy (for example, food). This energy is utilized for two purposes: mere maintenance and work. For man, maintenance of life requires about 1600 kilocalories a day ("maintenance calories"). Anything that he does over and above merely staying alive will be defined as work, and is supported by "work calories" which he takes in. Work calories are used not only for what we call work in common speech; they are also required for all forms of enjoyment, from swimming and automobile racing to playing music and writing poetry. If our goal is to maximize population it is obvious what we must do: We must make the work calories per person approach as close to zero as possible. No gourmet meals, no vacations, no sports, no music, no literature, no art. . . . I think that everyone will grant, without argument or proof, that maximizing population does not maximize goods. Bentham's goal is impossible.

In reaching this conclusion I have made the usual assumption that it is the acquisition of energy that is the problem. The appearance of atomic energy has led some to question this assumption. However, given an infinite source of energy, population growth still produces an inescapable problem. The problem of the acquisition of energy is replaced by the problem of its dissipation, as J. H. Fremlin has so wittily shown (4). The arithmetic signs in the analysis are, as it were, reversed; but Bentham's goal is still unobtainable.

The optimum population is, then, less than the maximum. The difficulty of defining the optimum is enormous; so far as I know, no one has seriously tackled this problem. Reaching an acceptable and stable solution will surely require more than one generation of hard analytical work—and much persuasion.

We want the maximum good per person; but what is good? To one person it is wilderness, to another it is ski lodges for thousands. To one it is estuaries to nourish ducks for hunters to shoot; to another it is factory land. Comparing one good with another is, we usually say, impossible because goods are incommensurable. Incommensurables cannot be compared.

Theoretically this may be true; but in real life incommensurables *are* commensurable. Only a criterion of judgment and a system of weighting are needed. In nature the criterion is survival. Is it better for a species to be small and hideable, or large and powerful? Natural selection commensurates the incommensurables. The compromise achieved depends on a natural weighting of the values of the variables.

Man must imitate this process. There is no doubt that in fact he already does, but unconsciously. It is when the hidden decisions are made explicit that the arguments begin. The problem for the years ahead is to work out an acceptable theory of weighting. Synergistic effects, nonlinear variation, and difficulties in discounting the future make the intellectual problem difficult, but not (in principle) insoluble.

Has any cultural group solved this practical problem at the present time, even on an intuitive level? One simple fact proves that none has: there is no prosperous population in the world today that has, and has had for some time, a growth rate of zero. Any people that has intuitively identified its optimum point will soon reach it, after which its growth rate becomes and remains zero.

Of course, a positive growth rate might be taken as evidence that a population is below its optimum. However, by any reasonable standards, the most rapidly growing populations on earth today are (in general) the most miserable. This association (which need not be invariable) casts doubt on the optimistic assumption that the positive growth rate of a population is evidence that it has yet to reach its optimum.

We can make little progress in working toward optimum poulation size until we explicitly exorcize the spirit of Adam Smith in the field of practical demography. In economic affairs, *The Wealth of Nations* (1776) popularized the "invisible hand," the idea that an individual who "intends only his own gain," is, as it were, "led by an invisible hand to promote . . . the public interest" (5). Adam Smith did not assert that this was invariably true, and perhaps neither did any of his followers. But he contributed to a dominant tendency of thought that has ever since interfered with positive action based on rational analysis, namely, the tendency to assume that decisions reached individually will, in fact, be the best decisions for an entire society. If this assumption is correct it justifies the continuance of our present policy of laissez-faire in reproduction. If it is correct we can assume that men will control their individual fecundity so as to produce the optimum population. If the assumption is not correct, we need to reexamine our individual freedoms to see which ones are defensible.

Tragedy of Freedom in a Commons

The rebuttal to the invisible hand in population control is to be found in a scenario first sketched in a little-known pamphlet (6) in 1833 by a mathematical amateur named William Forster Lloyd (1794–1852). We may well call it "the tragedy of the commons," using the word "tragedy" as the philosopher Whitehead used it (7): "The essence of dramatic tragedy is not unhappiness. It resides in the solemnity of the remorseless working of things." He then goes on to say, "This inevitableness of destiny can only be illustrated in terms of human life by incidents which in fact involve unhappiness. For it is only by them that the futility of escape can be made evident in the drama."

The tragedy of the commons develops in this way. Picture a pasture open to all. It is to be expected that each herdsman will try to keep as many cattle as possible on the commons. Such an arrangement may work reasonably satisfactorily for centuries because tribal wars, poaching, and disease keep the numbers of both man and beast well below the carrying capacity of the land. Finally, however, comes the day of reckoning, that is, the day when the long-desired goal of social stability becomes a reality. At this point, the inherent logic of the commons remorselessly generates tragedy.

As a rational being, each herdsman seeks to maximize his gain. Explicitly or implicitly, more or less consciously, he asks, "What is the utility *to me* of adding one more animal to my herd?" This utility has one negative and one positive component.

1) The positive component is a function of the increment of one animal. Since the herdsman receives all the proceeds from the sale of the additional animal, the positive utility is nearly +1

2) The negative component is a function of the additional overgrazing created by one more animal. Since, however, the effects of overgrazing are shared by all the herdsmen, the negative utility for any particular decision-making herdsman is only a fraction of −1.

Adding together the component partial utilities, the rational herdsman concludes that the only sensible course for him to pursue is to add another animal to his herd. And another; and another. . . . But this is the conclusion reached by each and every rational herdsman sharing a commons. Therein is the tragedy. Each man is locked into a system that compels him to increase his herd without limit—in a world that is limited. Ruin is the destination toward which all men rush, each pursuing his own best interest in a society that believes in the freedom of the commons. Freedom in a commons brings ruin to all.

Some would say that this is a platitude. Would that it were! In a sense, it was learned thousands of years ago, but natural selection favors the forces of psychological denial (8). The individual benefits as an individual from his ability to deny the truth even though society as a whole, of which he is a part, suffers. Education can counteract the natural tendency to do the wrong thing, but the inexorable succession of generations requires that the basis for this knowledge be constantly refreshed.

A simple incident that occurred a few years ago in Leominster, Massachusetts, shows how perishable the knowledge is. During the Christmas shopping season the parking meters downtown were covered with plastic bags that bore tags reading: "Do not open until after Christmas. Free parking courtesy of the mayor and city council." In other words, facing the prospect of an increased demand for already scarce space, the city fathers reinstituted the system of the commons. (Cynically, we suspect that they gained more votes than they lost by this retrogressive act.)

In an approximate way, the logic of the commons has been understood for a long time, perhaps since the discovery of agriculture or the invention of private property in real estate. But it is understood mostly only in special cases which are not sufficiently generalized. Even at this late date, cattlemen leasing national land on the western ranges demonstrate no more than an ambivalent understanding, in constantly pressuring federal authorities to increase the head count to the point where overgrazing produces erosion and weed-dominance. Likewise, the oceans of the world continue to suffer from the survival of the philosophy of the commons. Maritime nations still respond automatically to the shibboleth of the "freedom of the seas." Professing to believe in the "inexhaustible resources of the oceans," they bring species after species of fish and whales closer to extinction (9).

The National Parks present another instance of the working out of the tragedy of the commons. At present, they are open to all, without limit. The parks themselves are limited in extent—there is only one Yosemite Valley—whereas population seems to grow without limit. The values that visitors seek in the parks are steadily eroded. Plainly, we must soon cease to treat the parks as commons or they will be of no value to anyone.

What shall we do? We have several options. We might sell them off as private property. We might keep them as public property, but allocate the right to enter them. The allocation might be on the basis of wealth, by the use of an auction system. It might be on the basis of merit, as defined by some agreed-upon standards. It might be by lottery. Or it might be on a first-come, first-served basis, administered to long queues. These, I think, are all the reasonable possibilities. They are all objectionable. But we must choose—or acquiesce in the destruction of the commons that we call our National Parks.

Pollution

In a reverse way, the tragedy of the commons reappears in problems of pollution. Here it is not a question of taking something out of the commons, but of putting something in—sewage, or chemical, radioactive, and heat wastes into water; noxious and dangerous fumes into the air; and distracting and unpleasant advertising signs into the line of sight. The calculations of utility are much the same as before. The rational man finds that his share of the cost of the wastes he discharges into the commons is less than the cost of purifying his wastes before releasing them. Since this is true for everyone, we are locked into a system of "fouling our own nest," so long as we behave only as independent, rational, free-enterprisers.

The tragedy of the commons as a food basket is averted by private property, or something formally like it. But the air and waters surrounding us cannot readily be fenced, and so the tragedy of the commons as a cesspool must be prevented by different means, by coercive laws or taxing devices that make it cheaper for the polluter to treat his pollutants than to discharge them untreated. We have not progressed as far with the solution of this problem as we have with the first. Indeed, our particular concept of private property, which deters us from exhausting the positive resources of the earth, favors pollution. The owner of a factory on the bank of a stream—whose property extends to the middle of the stream—often has difficulty seeing why it is not his natural right to muddy the waters flowing past his door. The law, always behind the times, requires elaborate stitching and fitting to adapt it to this newly perceived aspect of the commons.

The pollution problem is a consequence of population. It did not much matter how a lonely American frontiersman disposed of his waste. "Flowing water purifies itself every 10 miles," my grandfather used to say, and the myth was near enough to the truth when he was a boy, for there were not too many people. But as population became denser, the natural chemical and biological recycling processes became overloaded, calling for a redefinition of property rights.

How To Legislate Temperance?

Analysis of the pollution problem as a function of population density uncovers a not generally recognized principle of morality, namely: *the morality of an act is a function of the state of the system at the time it is performed* (10). Using the commons as a cesspool does not harm the general public under frontier conditions, because there is no public; the same behavior in a metropolis is unbearable. A hundred and fifty years ago a plainsman could kill an American bison, cut out only the tongue for his dinner, and discard the rest of the animal. He was not in any important sense being wasteful. Today, with only a few thousand bison left, we would be appalled at such behavior.

In passing, it is worth noting that the morality of an act cannot be determined from a photograph. One does not know whether a man killing an elephant or setting fire to the grassland is harming others until one knows the total system in which his act appears. "One picture is worth a thousand words," said an ancient Chinese; but it may take 10,000 words to validate it. It is as tempting to ecologists as it is to reformers in general to try to persuade others by way of the photographic shortcut. But the essense of an argument cannot be photographed: it must be presented rationally —in words.

That morality is system-sensitive escaped the attention of most codifiers of ethics in the past. "Thou shalt not . . ." is the form of traditional ethical directives which make no allowance for particular circumstances. The laws of our society follow the pattern of ancient ethics, and therefore are poorly suited to governing a complex, crowded, changeable world. Our epicyclic solution is to augment statutory law with administrative law. Since it is practically impossible to spell out all the conditions under which it is safe to burn trash in the back yard or to run an automobile without smog-control, by law we delegate the details to bureaus. The result is administrative law, which is rightly feared for an ancient reason—*Quis custodiet ipsos custodes?*—"Who shall watch the watchers themselves?" John Adams said that we must have "a government of laws and not men." Bureau administrators, trying to evaluate the morality of acts in the total system, are singularly liable to corruption, producing a government by men, not laws.

Prohibition is easy to legislate (though not necessarily to enforce); but how do we legislate temperance? Experience indicates that it can be accomplished best through the mediation of administrative law. We limit possibilities unnecessarily if we suppose that the sentiment of *Quis custodiet* denies us the use of administrative law. We should rather retain the phrase as a perpetual reminder of fearful dangers we cannot avoid. The great challenge facing us now is to invent the corrective feedbacks that are needed to keep custodians honest. We must find ways to legitimate the needed authority of both the custodians and the corrective feedbacks.

Freedom To Breed Is Intolerable

The tragedy of the commons is involved in population problems in another way. In a world governed solely by the principle of "dog eat dog"—if indeed there ever was such a world—how many children a family had would not be a matter of public concern. Parents who bred too exuberantly would leave fewer descendants, not more, because they would be unable to care adequately for their children. David Lack and others have found that such a negative feedback demonstrably controls the fecundity of birds (*11*). But men are not birds, and have not acted like them for millenniums, at least.

If each human family were dependent only on its own resources; *if* the children of improvident parents starved to death; *if*, thus, overbreeding brought its own "punishment" to the germ line—*then* there would be no public interest in controlling the breeding of families. But our society is deeply committed to the welfare state (*12*), and hence is confronted with another aspect of the tragedy of the commons.

In a welfare state, how shall we deal with the family, the religion, the race, or the class (or indeed any distinguishable and cohesive group) that adopts overbreeding as a policy to secure its own aggrandizement (*13*)? To couple the concept of freedom to breed with the belief that everyone born has an equal right to the commons is to lock the world into a tragic course of action.

Unfortunately this is just the course of action that is being pursued by the United Nations. In late 1967, some 30 nations agreed to the following (*14*):

> The Universal Declaration of Human Rights describes the family as the natural and fundamental unit of society. It follows that any choice and decision with regard to the size of the family must irrevocably rest with the family itself, and cannot be made by anyone else.

It is painful to have to deny categorically the validity of this right; denying it, one feels as uncomfortable as a resident of Salem, Massachusetts, who denied the reality of witches in the 17th century. At the present time, in liberal quarters, something like a taboo acts to inhibit criticism of the United Nations. There is a feeling that the United Nations is "our last and best hope," that we shouldn't find fault with it; we shouldn't play into the hands of the archconservatives. However, let us not forget what Robert Louis Stevenson said: "The truth that is suppressed by friends is the readiest weapon of the enemy." If we love the truth we must openly deny the validity of the Universal Declaration of Human Rights, even though it is promoted by the United Nations. We should also join with Kingsley Davis (*15*) in attempting to get Planned Parenthood-World Population to see the error of its ways in embracing the same tragic ideal.

Conscience Is Self-Eliminating

It is a mistake to think that we can control the breeding of mankind in the long run by an appeal to conscience. Charles Galton Darwin made this point when he spoke on the centennial of the publication of his grandfather's great book. The argument is straightforward and Darwinian.

People vary. Confronted with appeals to limit breeding, some people will undoubtedly respond to the plea more than others. Those who have more children will produce a larger fraction of the next generation than those with more susceptible consciences. The difference will be accentuated, generation by generation.

In C. G. Darwin's words: "It may well be that it would take hundreds of generations for the progenitive instinct to develop in this way, but if it should do so, nature would have taken her revenge, and the variety *Homo contracipiens* would become extinct and would be replaced by the variety *Homo progenitivus*" (16).

The argument assumes that conscience or the desire for children (no matter which) is hereditary—but hereditary only in the most general formal sense. The result will be the same whether the attitude is transmitted through germ cells, or exosomatically, to use A. J. Lotka's term. (If one denies the latter possibility as well as the former, then what's the point of education?) The argument has here been stated in the context of the population problem, but it applies equally well to any instance in which society appeals to an individual exploiting a commons to restrain himself for the general good—by means of his conscience. To make such an appeal is to set up a selective system that works toward the elimination of conscience from the race.

Pathogenic Effects of Conscience

The long-term disadvantage of an appeal to conscience should be enough to condemn it; but has serious short-term disadvantages as well. If we ask a man who is exploiting a commons to desist "in the name of conscience," what are we saying to him? What does he hear?—not only at the moment but also in the wee small hours of the night when, half asleep, he remembers not merely the words we used but also the nonverbal communication cues we gave him unawares? Sooner or later, consciously or subconsciously, he senses that he has received two communications, and that they are contradictory: (i) (intended communication) "If you don't do as we ask, we will openly condemn you for not acting like a responsible citizen"; (ii) (the unintended communication) "If you *do* behave as we ask, we will secretly condemn you for a simpleton who can be shamed into standing aside while the rest of us exploit the commons."

Everyman then is caught in what Bateson has called a "double bind." Bateson and his co-workers have made a plausible case for viewing the double bind as an important causative factor in the genesis of schizophrenia (17). The double bind may not always be so damaging, but it always endangers the mental health of anyone to whom it is applied. "A bad conscience," said Nietzsche, "is a kind of illness."

To conjure up a conscience in others is tempting to anyone who wishes to extend his control beyond the legal limits. Leaders at the highest level succumb to this temptation. Has any President during the past generation failed to call on labor unions to moderate voluntarily their demands for higher wages, or to steel companies to honor voluntary guidelines on prices? I can recall none. The rhetoric used on such occasions is designed to produce feel-

110

ings of guilt in noncooperators.

For centuries it was assumed without proof that guilt was a valuable, perhaps even an indispensable, ingredient of the civilized life. Now, in this post-Freudian world, we doubt it.

Paul Goodman speaks from the modern point of view when he says: "No good has ever come from feeling guilty, neither intelligence, policy, nor compassion. The guilty do not pay attention to the object but only to themselves, and not even to their own interests, which might make sense, but to their anxieties" (18).

One does not have to be a professional psychiatrist to see the consequences of anxiety. We in the Western world are just emerging from a dreadful two-centuries-long Dark Ages of Eros that was sustained partly by prohibition laws, but perhaps more effectively by the anxiety-generating mechanisms of education. Alex Comfort has told the story well in *The Anxiety Makers* (19); it is not a pretty one.

Since proof is difficult, we may even concede that the results of anxiety may sometimes, from certain points of view, be desirable. The larger question we should ask is whether, as a matter of policy, we should ever encourage the use of a technique the tendency (if not the intention) of which is psychologically pathogenic. We hear much talk these days of responsible parenthood; the coupled words are incorporated into the titles of some organizations devoted to birth control. Some people have proposed massive propaganda campaigns to instill responsibility into the nation's (or the world's) breeders. But what is the meaning of the word responsibility in this context? Is it not merely a synonym for the word conscience? When we use the word responsibility in the absence of substantial sanctions are we not trying to browbeat a free man in a commons into acting against his own interest? Responsibility is a verbal counterfeit for a substantial *quid pro quo*. It is an attempt to get something for nothing.

If the word responsibility is to be used at all, I suggest that it be in the sense Charles Frankel uses it (20). "Responsibility," says this philosopher, "is the product of definite social arrangements." Notice that Frankel calls for social arrangements—not propaganda.

Mutual Coercion

Mutually Agreed upon

The social arrangements that produce responsibility are arrangements that create coercion, of some sort. Consider bank-robbing. The man who takes money from a bank acts as if the bank were a commons. How do we prevent such action? Certainly not by trying to control his behavior solely by a verbal appeal to his sense of responsibility. Rather than rely on propaganda we follow Frankel's lead and insist that a bank is not a commons; we seek the definite social arrangements that will keep it from becoming a commons. That we thereby infringe on the freedom of would-be robbers we neither deny nor regret.

The morality of bank-robbing is particularly easy to understand because we accept complete prohibition of this activity. We are willing to say "Thou shalt not rob banks," without providing for exceptions. But temperance also can be created by coercion. Taxing is a good coercive device. To keep downtown shoppers temperate in their use of parking space we introduce parking meters for short periods, and traffic fines for longer ones. We need not actually forbid a citizen to park as long as he wants to; we need merely make it increasingly expensive for him to do so. Not prohibition, but carefully biased options are what we offer him. A Madison Avenue man might call this persuasion; I prefer the greater candor of the word coercion.

Coercion is a dirty word to most liberals now, but it need not forever be so. As with the four-letter words, its dirtiness can be cleansed away by exposure to the light, by saying it over and over without apology or embarrassment. To many, the word coercion implies arbitrary decisions of distant and irresponsible bureaucrats; but this is not a necessary part of its meaning. The only kind of coercion I recommend is mutual coercion, mutually agreed upon by the majority of the people affected.

To say that we mutually agree to coercion is not to say that we are required to enjoy it, or even to pretend we enjoy it. Who enjoys taxes? We all grumble about them. But we accept compulsory taxes because we recognize that voluntary taxes would favor the conscienceless. We institute and (grumblingly) support taxes and other coercive devices to escape the horror of the commons.

An alternative to the commons need not be perfectly just to be preferable. With real estate and other material goods, the alternative we have chosen is the institution of private property coupled with legal inheritance. Is this system perfectly just? As a genetically trained biologist I deny that it is. It seems to me that, if there are to be differences in individual inheritance, legal possession should be perfectly correlated with biological inheritance—that those who are biologically more fit to be the custodians of property and power should legally inherit more. But genetic recombination continually makes a mockery of the doctrine of "like father, like son" implicit in our laws of legal inheritance. An idiot can inherit millions, and a trust fund can keep his estate intact. We must admit that our legal system of private property plus inheritance is unjust—but we put up with it because we are not convinced, at the moment, that anyone has invented a better system. The alternative of the commons is too horrifying to contemplate. Injustice is preferable to total ruin.

It is one of the peculiarities of the warfare between reform and the status quo that it is thoughtlessly governed by a double standard. Whenever a reform measure is proposed it is often defeated when its opponents triumphantly discover a flaw in it. As Kingsley Davis has pointed out (21), worshippers of the status quo sometimes imply that no reform is possible without unanimous agreement, an implication contrary to historical fact. As nearly as I can make out, automatic rejection of proposed reforms is based on one of two unconscious assumptions: (i) that the status quo is perfect; or (ii) that the choice we face is between reform and no action; if the proposed reform is imperfect, we presumably should take no action at all, while we wait for a perfect proposal.

But we can never do nothing. That which we have done for thousands of years is also action. It also produces evils. Once we are aware that the status quo is action, we can then compare its discoverable advantages and disadvantages with the predicted advantages and disadvantages of the proposed reform, discounting as best we can for our lack of experience. On the basis of such a comparison, we can make a rational decision which will not involve the unworkable assumption that only perfect systems are tolerable.

Recognition of Necessity

Perhaps the simplest summary of this analysis of man's population problems is this: the commons, if justifiable at all, is justifiable only under conditions of low-population density. As the human population has increased, the commons has had to be abandoned in one aspect after another.

First we abandoned the commons in food gathering, enclosing farm land and restricting pastures and hunting and fishing areas. These restrictions are still not complete throughout the world.

Somewhat later we saw that the commons as a place for waste disposal would also have to be abandoned. Restrictions on the disposal of domestic sewage are widely accepted in the Western world; we are still struggling to close the commons to pollution by automobiles, factories, insecticide sprayers, fertilizing operations, and atomic energy installations.

In a still more embryonic state is our recognition of the evils of the commons in matters of pleasure. There is almost no restriction on the propagation of sound waves in the public medium. The shopping public is assaulted with mindless music, without its consent. Our government is paying out billions of dollars to create supersonic transport which will disturb 50,000 people for every one person who is whisked from coast to coast 3 hours faster. Advertisers muddy the airwaves of radio and television and pollute the view of travelers. We are a long way from outlawing the commons in matters of pleasure. Is this because our Puritan inheritance makes us view pleasure as

something of a sin, and pain (that is, the pollution of advertising) as the sign of virtue?

Every new enclosure of the commons involves the infringement of somebody's personal liberty. Infringements made in the distant past are accepted because no contemporary complains of a loss. It is the newly proposed infringements that we vigorously oppose; cries of "rights" and "freedom" fill the air. But what does "freedom" mean? When men mutually agreed to pass laws against robbing, mankind became more free, not less so. Individuals locked into the logic of the commons are free only to bring on universal ruin; once they see the necessity of mutual coercion, they become free to pursue other goals. I believe it was Hegel who said, "Freedom is the recognition of necessity."

The most important aspect of necessity that we must now recognize, is the necessity of abandoning the commons in breeding. No technical solution can rescue us from the misery of overpopulation. Freedom to breed will bring ruin to all. At the moment, to avoid hard decisions many of us are tempted to propagandize for conscience and responsible parenthood. The temptation must be resisted, because an appeal to independently acting consciences selects for the disappearance of all conscience in the long run, and an increase in anxiety in the short.

The only way we can preserve and nurture other and more precious freedoms is by relinquishing the freedom to breed, and that very soon. "Freedom is the recognition of necessity"— and it is the role of education to reveal to all the necessity of abandoning the freedom to breed. Only so, can we put an end to this aspect of the tragedy of the commons.

References

1. J. B. Wiesner and H. F. York, *Sci. Amer.* 211 (No. 4), 27 (1964).
2. G. Hardin, *J. Hered.* 50, 68 (1959); S. von Hoernor. *Science* 137, 18 (1962).
3. J. von Neumann and O. Morgenstern, *Theory of Games and Economic Behavior* (Princeton Univ. Press, Princeton, N.J., 1947), p. 11.
4. J. H. Fremlin. *New Sci.*, No. 415 (1964), p. 285.
5. A. Smith. *The Wealth of Nations* (Modern Library, New York, 1937), p. 423.
6. W. F. Lloyd. *Two Lectures on the Checks to Population* (Oxford Univ. Press, Oxford, England. 1833), reprinted (in part) in *Population, Evolution, and Birth Control*, G. Hardin, Ed. (Freeman. San Francisco, 1964), p. 37.
7. A. N. Whitehead, *Science and the Modern World* (Mentor, New York, 1948), p. 17.
8. G. Hardin, Ed. *Population, Evolution, and Birth Control* (Freeman, San Francisco, 1964), p. 56.
9. S. McVay, *Sci. Amer.* 216 (No. 8), 13 (1966).
10. J. Fletcher, *Situation Ethics* (Westminster, Philadelphia, 1966).
11. D. Lack, *The Natural Regulation of Animal Numbers* (Clarendon Press, Oxford, 1954).
12. H. Girvetz, *From Wealth to Welfare* (Stanford Univ. Press, Stanford, Calif., 1950).
13. G. Hardin, *Perspec. Biol. Med.* 6, 366 (1963).
14. U. Thant, *Int. Planned Parenthood News*, No. 168 (February 1968), p. 3.
15. K. Davis, *Science* 158, 730 (1967).
16. S. Tax, Ed., *Evolution after Darwin* (Univ. of Chicago Press, Chicago, 1960), vol. 2, p. 469.
17. G. Bateson, D. D. Jackson, J. Haley, J. Weakland, *Behav. Sci.* 1, 251 (1956).
18. P. Goodman, *New York Rev. Books* 10(8), 22 (23 May 1968).
19. A. Comfort, *The Anxiety Makers* (Nelson, London, 1967).
20. C. Frankel, *The Case for Modern Man* (Harper, New York, 1955), p. 203.
21. J. D. Roslansky, *Genetics and the Future of Man* (Appleton-Century-Crofts, New York, 1966), p. 177.

114

MAN'S PARTICIPATORY EVOLUTION

IS MAN OVER ADAPTING TO HIS ENVIRONMENT?

Mr. Dubos, who is member of the Rockefeller University faculty, will become director of environmental studies at the new State University of New York College at Purchase. His latest book is Reason Awake: Science for Man *(See* Current, *June 1970, page 36). The following article "Man Over Adapting" by* René Dubos *is from* Psychology Today, *February 1971.*

René Dubos

Until the past few decades, most technological and social changes were gradual and affected only a small percentage of the population at any given time. The rate of change was slow enough for man to adapt—the physiological and anatomical characteristics of his body underwent alterations to fit the new circumstances, and so did his mental attitudes and social structures.

But now the environment is changing so rapidly that the processes of biological, mental and social adaptation cannot keep pace. A tragedy of modern life is that the experience of the father is of little use to his children.

As everyone knows, man's life expectancy at birth has greatly increased during recent decades, especially in affluent groups. This has been due almost entirely to a reduction in infant mortality. People do not live significantly longer than they used to, but medical advances have allowed more of them to make it through nutritional and microbial hazards of infancy.

Although life expectancy at birth has increased, the expectancy of life past the age of 45 has not changed significantly, if at all. Modern adults do not live longer than their counterparts at the beginning of the Century. Even with affluence and modern medical care, they are still disease-ridden. Cardiac conditions, cerebral strokes, various types of cancers, arthritis, emphysema, bronchitis, and mental afflictions are among the many chronic ailments that plague all affluent technological societies—they are the diseases of civilization.

Death by these diseases is not due to lack of medical care. In the United States, for example, scientists and especially physicians have shorter life expectancies than members of other social groups even though they belong to privileged social classes that have ready access to medical attention.

The simple fact is that we know very little about most

chronic and degenerative diseases. We do know, however, that they are not inherent in man's nature, but are caused by environmental and social influences that are ubiquitous in the technological world. They are the expressions of man's failure to respond successfully to modern ways of life.

An analogy may be seen in the first phase of the Industrial Revolution during the 18th Century. Many of the workers had come recently from agricultural areas. They found it difficult to adapt to the appalling conditions in the cities, and there was a profound deterioration in their physical and mental health. Later, through a multiplicity of medical and social measures men were able to adapt to the health threats associated with the factories and tenements in mushrooming industrial cities. But new health problems arise from any sudden and profound disturbance in man's way of life.

Our limited adaptive potential Clearly our adaptive potentialities are not unlimited. Even now they may be exceeded by some of the stresses created by contemporary technological developments. In the course of his prehistoric evolution, man was repeatedly exposed to seasonal famine, inclement weather, infectious processes, physical fatigue, and many forms of fear. This evolutionary experience has generated in his genetic constitution the potentiality to adapt to many different kinds of stresses.

But he now faces dangers that have no precedent in his evolutionary past. He probably does not possess the responses that will be necessary to adapt to many of the new environmental threats created by modern technology: the toxic effects of chemical pollutants and synthetic substances, the physiological and mental aberrations resulting from the mechanizations of life, the artificial and violent stimuli that are ubiquitous in the technological world.

Modern life is almost completely divorced from the cosmic cycles under which man evolved and to which his constitution is intimately geared. Every person shifted from day to night duty or vice versa is aware of the mental and physical difficulties that result from disturbances in his hormonal rhythms. Travelers who jet from one continent to another have experienced discomforts that may last for days—until their bodies adjust to the new cycles. Perhaps these physiological disturbances will have to be paid for later in the form of pathological effects—we do not know yet, because the phenomenon is so new.

We know that chickens will develop pathological symptoms if they are exposed throughout the year to artificial light for increasing egg production, but hardly anything is known of the effects produced in man by disturbances of his normal physiological rhythms.

In brief, there is no doubt that man is still immensely adaptable, but it is also certain he cannot adapt to just anything and everything. And the rate of his biological evolution is so slow that it cannot possibly keep pace with the rate of the technological and social change.

I don't deny that man has a remarkable capacity to adapt to new conditions. He has survived the horrible ordeals of modern warfare; he has multiplied in crowded and polluted cities; and he has worked effectively in atmospheres clouded by tobacco smoke and chemical fumes amid the infernal noise of telephones, typewriters and loud mechanical contrivances.

Because man adapted during evolutionary times to the vicissitudes of the Stone Age, it is often assumed that he can adapt now to all the traumatic conditions created by modern technology. But this is improbable. Conditions now exist that are without precedent in the biological history of man, and the chances are small that he is equipped to adapt to all of them.

Nutrition and modern life

Immense progress has been made during the past century in the theory of nutrition and in the practical aspects of food production. But unfortunately, little is known of the kind of nutrition best suited to modern urban life. Nutritional requirements were determined two generations ago for vigorous and physically active young men, but these requirements do not necessarily fit automated, air-conditioned life. A new paradox is that the appetites our ancestors evolved to meet certain physiological needs may now be dangerous because strenuous physical exercise is no longer a part of everyday life.

Much remains to be learned about the nutritional needs of pregnant women and infants. The biggest baby is not necessarily the one that will be healthiest as an adult. Too generous a diet during early life may so imprint the child that his nutritional demands remain excessively large thereafter —deleterious effects in the long run.

Environmental pollution provides another example of man's ability to function in a biologically undesirable environment, and also of the dangers inherent in this adaptability.

Since the beginning of the Industrial Revolution, inhabitants of Northern Europe have been exposed to a variety of air pollutants produced by coal and by the fumes from chemical plants, which are rendered even more objectionable by the inclemency of the Atlantic climate. But after long experience with pollution and with bad weather, Northern Europeans have developed physiological reactions and living habits that have adaptive value—they accept their dismal environment almost cheerfully. But even in those who seem to have adapted to irritating atmosphere, the respiratory system continuously registers the insult of air pollutants. As a result chronic pulmonary disease is now the greatest single medical problem in Great Britain. It is increasing at an alarming rate in North America, and it will probably spread to all areas undergoing industrialization.

There is evidence that air pollution increases the incidence of various types of cancers as well as the fatalities among persons suffering from vascular diseases.

The delayed effects of air pollutants constitute models for the kind of medical problems that are likely to arise in the future from other forms of environmental pollution. People will insist that chemical pollution of air, water and food be sufficiently controlled to prevent immediately disabling and obvious toxic effects. But they will tolerate milder concentrations of environmental pollutants that do not interfere with social and economic life. Continued exposure to low levels of toxic agents will eventually result in a great variety of delayed pathology that will *not* be detected at the time of exposure and may not become evident *until several decades later.*

Noise is another aspect of environmental pollution that certain human beings come to tolerate, but at great cost. People adapt to continuous exposure to loud and painful noise by shutting out the objectionable sounds from perception. This, however, does not prevent destructive anatomical effects from taking place. There may be an impairment of hearing— a permanent inability to hear certain frequencies. The cost of adaptation to noise is therefore a loss in the enjoyment of music and of the more subtle qualities of the human voice.

Adaptation to crowding may also have unfortunate results in the long run. Admittedly, man is a gregarious animal who commonly seeks crowded environments. But this does not mean that man can indefinitely increase the density of his populations; it means only that the safe limits are not known. In animals, crowding beyond a certain level results in behavioral and even physiological disturbances. Man has generally avoided the worst of these disturbances through a variety of social and architectural conventions and especially by learning to develop psychological unawareness of his surroundings. In extremely crowded environments each of us lives—as it were—in a world of his own. But eventually this adaptation to crowding decreases man's ability to relate to other human beings. He may become unaware of their presence and totally antisocial.

ADAPTATION OR SOCIAL CHANGE?

Throughout prehistory and history, man has proved his ability to make adjustments that tend to correct the disturbing effects of the environment. Such adaptive responses contribute to the welfare of the organism at the time they occur, but may be deleterious at a later date. Many of man's chronic disorders are the secondary and delayed consequences of homeostatic responses that were adaptive at first but are faulty in the long run. For example, the production of scar tissue is a homeostatic response because it heals wounds and helps in checking the spread of infection. But when scar tissue forms in the liver or kidney, it means cirrhosis or nephritis; scar tissue may freeze the joints in rheumatoid arthritis or may choke the breathing process in the lung. In other words, homeostatic processes may have an immediate protective or reparative function, but they can become destructive in the long run. Wisdom of the body is often very short-sighted wisdom.

Adaptation and mental health

Adaptation can also be dangerous for mental health. Man seems to be adapting to the ugliness of smoky skies, polluted streams, and anonymous buildings; to life without the fragrance of flowers, the song of birds, and other pleasurable stimuli from nature. This adaptation, however, is only superficial, and destructive in the long run. Air, water, earth, fire, the subtle forces of the cosmos, the natural rhythms and diversity of life have shaped man's nature during the evolutionary past and have created deep-rooted sensual and emotional needs that cannot be eradicated. The impoverishment of sensual and emotional life will progressively result in the atrophy of our uniquely human attributes. Like the giant Antaeus in the Greek legend, man loses his strength when he loses contact with the earth.

The greatest improvements in health during the past century have resulted from the continuous rise in our standards of living. But we may now be coming to the phase of diminishing returns. Our prosperity creates a new set of medical problems. Environmental pollution, excessive food intake, emotional deprivation, lack of physical exercise, the constant bombardment of unnatural stimuli, man's estrangement from natural biological rhythms—these are just some of the many consequences of urbanized and industrialized life that have direct or indirect pathological effects.

It can no longer be taken for granted that a further rise in living standards will continue to bring about improvements in health. More probably it will result in new patterns of diseases, it will prove extremely difficult to control them because all aspects of the urban and industrial environment are so intimately interwoven in the social fabric.

Men will accept collective measures to protect the environment. Keeping streets and houses clear of refuse, filtering and chlorinating the water supplies, watching over the purity of food products, assuring a minimum of safe air in public places—these measures can be applied collectively, anonymously, and without interfering seriously with individual freedom.

But any measure that requires individual discipline and personal effort is likely to be neglected. Almost everybody is aware of the dangers associated with overeating, failure to engage in physical exercise, chain cigarette smoking, excessive consumption of drugs, constant exposure to pollution, noise and people. But few persons are willing to make the individual efforts necessary to avoid these dangers. And the consequences of environmental threats are so often indirect and delayed that the public is hardly aware of them.

The mass diseases of the past were connected quite directly with the natural environment. But the chronic and degenerative diseases and the mental disorders of today are integrated in a much more complex way with the sociocultural environment. For this reason they are much less amenable to community-based control than are the nutritional and infectious diseases, and demand that greater emphasis be placed on the cooperation and interest of the individual person. It is therefore necessary to reformulate medical policies so that the public-health practices that emerged from 19th-Century science will be supplemented by more personal relationships between physician and patient.

Designing new environments

We shall of course develop new medical procedures for the treatment of the chronic and degenerative diseases—including the mental ones—that constitute the chief health problems of technological societies. We shall also develop protective technologies against environmental insults. But if we depend exclusively on such defense measures, we shall increasingly behave like hunted creatures, running from one therapeutic or protective device to another, each more complex and more costly than the one before. We shall end by spending much of our energy defending ourselves against environmental threats of our own creation, while sacrificing thereby the values that make life worth living.

Man should not try to conform to the environment created by social and technological innovations; he should instead design environments really adapted to his nature. He should not be satisfied, with palliative measures treating the effects of objectionable conditions, but instead change the conditions. Now that scientific technology has made us so powerful, and so destructive, we must try to imagine the kinds of surroundings and ways of life we desire, lest we end up with a jumble of technologies and countertechnologies that will eventually smother body and soul.

CONTEMPORARY PERSPECTIVE:
THE ORDERED TECHNOLOGY

Commitment to Humanity

by R. Buckminster Fuller

I often hear myself spoken of as a technologist. I chose various strategies in my life in order to be effective, and those strategies did bring me into technology. A great many people therefore think of me as being so vigorously concerned with technology that I lack humanist considerations. But my commitment has always been to humanity.

The recitation of my various undertakings seems to many people to be numerically impressive. But the list would not be even mildly surprising to one who had lived the particular kind of life that I have. Perhaps it would be useful to reflect upon some of the influences in my life and on some of the basic assumptions that emerged in my thought.

I should point out that I deliberately peeled off in 1927 from the patterns I had found to represent what most members of society felt were the bounds of social considerations. And because I deliberately took a new direction, I had a great deal of time at my disposal. In my life there has never been a moment for which I have been more grateful than that one. While it is true that I started in this new direction in 1927, I want to make it clear that even before then I was completely committed to humanity.

The Love and Life of a Child

An important influence in my life was the death of our first child. Born just as World War I was ending, our child first caught the flu, followed by infantile paralysis and spinal meningitis. She survived those things and lived until just before her fourth birthday. You can imagine the intense love we had for this wonderful darling who was physically in-

This article first appeared in THE HUMANIST, May/June 1970, and is reprinted by permission.

capacitated. The illnesses did not affect her brain, but she was unable to run around and gratify the drives and curiosities all new life has. She was forced to gain information through other people's motions and to use their senses.

She demonstrated extraordinary compensations of the Emersonian kind. Her sensitivity to what the people around her were thinking became astounding. Many, many times, as one of us had formulated a thought and were just about to speak, she would utter the words first. Frequently, the thought and the words involved would not really be within her ken. This certainly convinced me that a young life is born with a very great potential.

A Second Child

With the birth of our second child, I found myself doing the first really good thinking, thinking on my own, that I had ever done. I became convinced that the responsibility of this new child had to be very great. It was a fantastic responsibility, a new life coming to us. I said if this new child were to be the kind of child I hoped she would be, she would become very unhappy if, as she grew older, she found I devoted myself to trying to bring advantage to her, then found herself in a world where there was great disadvantage for others. But, on the other hand, she would be extremely happy if she discovered as she grew up that I had committed myself, not merely to solving problems for her, but I had committed myself to the attempt to solve humanity's problems.

By that time I had become convinced that the conditioned reflexes of my fellows, both my contemporaries and people older than myself, were so badly aimed, so miscued with regard to what seemed to me to be some of the verities, that I had no feeling I could ever be of any use to humanity by trying to persuade my fellow man to behave in ways different from the ways they were behaving.

I pondered a great deal on what the individual, operating on his own initiative, might be able to do on behalf of his fellow man. I wondered if there might not be ways in which he could be more effective than massive corporations and massive states. I searched for strategies which might be employed to those ends. I decided there were two questions involved. First, how can an individual function? Second, what would be an individual's highest priority? It seemed quite clear that new life has an extraordinarily high potential, and any way in which I could modify the environment so that new life and all the new potentials might prosper would be worthwhile. I was confident that all life is born genius, and simply gets degeniused very rapidly by circumstances.

123

Modifying the Environment

It then became very clear to me that I should be concerned with reforming the environment and not with reforming human beings. By environment I mean everything that is in me, not just some things and not just static objects, but the behavior of all nature including, particularly, human beings. I decided to rearrange the scenery; and I became interested in ways in which nature permits the scenery to be rearranged.

I became thoroughly convinced that the phenomenon of entropy, in which all local systems lose energy, meant that every local system in giving off its energy gave it off to the environment and therefore ordered the environment. All local systems are continually generating change and have periodicity. Local systems all have patterns that do not correspond with other systems, and they are unique. While each is regular and orderly as it gives off its energies, these do not necessarily mesh, and they seem to be disorderly with regard to the rest of the system.

The Balance of Nature and Generalized Principles

I began to search for what might balance this, because I became convinced that nature does balance everything. Quite clearly the stars are giving off energies in disorderly rays and that's how we are able to see them and their radiation. I thought it possible that our own world and the work of man aboard our particular planet were gathering energies and that energies are arriving here in the form of cosmic radiation and energy from the sun and being impounded. We find vegetables and algae impounding energy. Biological systems are producing by photosynthesis orderly molecules that are clearly anti-entropic. Thus, a form of order develops. But I thought that by far the most powerful order that we know is the ability of the human mind to sort and discover principles.

There has been considerable confusion over the words "generalized principles." In literature, when a man speaks about a generalization, it means the speaker is covering too much territory too thinly to be persuasive. In science, a generalization is the statement of behavior that has been discovered to be operative and that holds true in every special case without exception. The human mind can detect this principle to be operative and holding true in every special case quite independent of the materials, and seemingly in very different circumstances.

I will give a very simple illustration in terms of the principle of leverage: we can review very quickly how someone first discovered leverage. Nobody told him about it. You go through a woods. You find that due to a storm trees have fallen before you. You want to go in as direct a line as you can, so you find yourself climbing over some of the trees. One of the trees you are climbing over suddenly begins to go down slowly. You feel the sinking and retreat from it, and the tree rises again. You discover that this enormous tree is lying across another tree and for that reason lifts easily. That is the beginning of scientific generalization, because not only could it be any tree but it could have been a steel bar or a column of reinforced concrete. The mathematical principle concerns the distance from the fulcrum and the amount of leverage in respect to the distance the log is lifted.

I find that only human minds seem to have this kind of generalizing capability. The brain is always dealing with specialized cases; it is a system of storing and retrieving special information. What is unique about the mind, it seems to me, is the absolutely weightless capability to survey experiences and discover a generalized principle operative. But having discovered a generalized principle, you find that you cannot design a generalized pattern yourself, even though you understand it. You can only design special cases. The physical side is always going to be the special case, and the metaphysical side is always going to be the generalization—complete, abstract, weightless. This seemed to me a very powerful point.

Education for Specialization

While at Harvard, I learned from my friends and all my reading and studying that at all of the universities in the land there was an increasing trend to specialization. Alfred North Whitehead wrote about the specialization at Harvard and other universities. He pointed out that in Europe a graduate scholar could go on with his work and find the authorities and the right books on various subjects that he would like to study. You simply had to find the right man. But Harvard was the first university to inaugurate completely separate graduate schools. Whitehead said it was because Americans liked the idea of specialists and believed that specialists meant champions and a very powerful team, and this would mean that society would prosper. But Whitehead also pointed out that the graduate school with its completely separate buildings and staff was a very expensive undertaking, and it became very much isolated from the rest of the university activity.

The nuances of specialization multiplied very rapidly. Bright individuals were persuaded to go into the graduate schools by examination and sifting. And these bright ones are now arriving at the graduate schools with a much finer specialization, with all energies going into a linear acceleration in that specific direction. So, like rockets, they get very far out. Maybe it would be like stars, but like stars they would be very remote from one another.

Whitehead also pointed out that these specialists did not find any spontaneous way in which they could communicate about their specialization. So they all talked about baseball. Because they were not able to put together their high potential, and society was expecting some harvesting of the potential, it had to be left to others to integrate their capabilities. So, Whitehead said, having deliberately sifted out the bright ones and making them all specialists who couldn't put their work together, you had to leave it to the dull ones to put things together. I call this Whitehead's Dilemma.

The great corporations tended to specialize. Automobile corporations produced an automobile. They recognized that it could not run across an open field. A roadway is part of an automobile, but is too complex a part of an automobile for them to produce. What they did was to manufacture a very attractive automobile and tantalize society, so that they would want roads. But they left it to the politicians, the dropouts who were not capable of specialization, to build the roads. The politicians saw that the people wanted roadways in order to be able to realize the automobile and enjoy it, and so they simply produced fantastic amounts of roadway in order to be elected. In fact the larger and more comprehensive the undertaking, the more it tends to be left to lower and lower echelons to coordinate matters. This also seems to be the case in international affairs, where the most talented don't often operate.

Whitehead pointed out that because Harvard specialized, other private and public schools immediately felt that they had to do it also. So, specialization and the graduate school idea became rampant throughout society.

Training of the Comprehensivist

I was astonished to find exactly the opposite was operative at the Naval Academy. They deliberately sorted their students, and they deliberately set about to make comprehensivists of the brightest. And they did this for a very fundamental reason—the Navy had to be concerned with the whole world or nothing, for three-fourths of the earth is water. Up to and including World War I, when I was in the Navy, men were being prepared to operate the most powerful tool in the world, the battleship. It incorporated everything that man had learned in chemistry, physics and mathematics about generalized principles, and every general case wherein you could arrange to carry the greatest hitting power for the greatest distance with the greatest accuracy and the least effort. You could float a fantastic amount of technology that you couldn't possibly move around on land.

I saw that the individual being prepared to take command of a battleship and operate it with his own common sense was going to have to operate autonomously—there was not going to be any contact with central authority. He had to know how to handle thousands and thousands of men, keeping them in good health, well-disciplined, and trained; he had to know how to build, and how to anticipate in a very long way because this was very long-distance thinking. He had to understand what the wishes of his society were; he had to understand what the wishes were of those who were running the world; for the battleship, in its day, was the basis of world power and strategy.

I became fascinated with the fact that at the Naval Academy they were training their men to be *comprehensivists*— to understand the world, to understand technology, etc. The Navy was able to take a ton and throw it and hit a ship over the horizon—and on the first throw. This required a high level of organized capability. I was at the Naval Academy at a very extraordinary moment, during the last days of the organized development of the comprehensivist. They were the extraordinary days in which the mastery of the earth was held by the British Navy, the American Navy then being a second-rate Navy which had not yet achieved parity. After World War I, we learned to scramble messages electronically, and at this point authority was centralized. The Navy started specializing after that, and it began to develop the submarine and the naval aviator, etc. I was among the handful of young men at the Naval Academy at the time when those who were running the world were having to educate young men to eventually take this generalized type of capability. If you were really interested and wanted to listen and find out, you realized you were being taught how to run the world. You were learning to look at things in a big way.

I think I had very powerful thoughts while in the Navy. I thought about the kind of responsibilities that I had, particularly when I was the skipper of smaller ships, and the fact that I was trying to understand technology in larger terms— ballistics, theories of ships, navigation, even large patterns of commerce and industry. I was interested here in the impressive harvesting of science that had gone into the extraordinary naval equipment; and I was one of the young human beings trained to understand that equipment and understand the bigger patterns. Yet I thought that something was missing here. What I recognized as wrong was the assumption that enormous resources and capability must be used destructively. Clearly the assumption goes back to statecraft and to Thomas Malthus.

Until Malthus, all the great empires of man—Genghis Khan, Alexander the Great, Julius Caesar—existed when men thought the earth was infinitely extensive. The empire was civilization and outside of the empire you kept some very dangerous people. Beyond the empire there were only dragons. The British Empire was the first empire of man to arise after man realized that he was living on a sphere.

The sphere was a closed system and not an open system. In an open system such as the Roman Empire there were an infinite number of variables. If you didn't like the way things were going, there were an infinite number of possibilities. With the closed system, however, we have Thomas Malthus, the first man to derive the total vital statistics of the closed system. From these statistics Malthus found that people were reproducing themselves much more rapidly than producing goods to support themselves. They were reproducing themselves at a geometrical rate and only producing goods at an arithmetical rate. Therefore, it was suddenly disclosed that man was designed to be a failure. There would never be enough for him. It was an horrendous kind of fact. Thirty-five years after Malthus we have the scientists, the geologists, and the biologists being taken around this closed system by men whom I call the great outlaws, because they lived outside the law. They wrote the law and theirs was the only law there was, and they made it as they went along. Because they realized that scientists could see things they could not, these great outlaws, who were really masters, took their scientists around the world to see what the resources were that could be developed. That's how Darwin made his trip around the world on *The Beagle*.

Darwin could not have developed the theory of evolution in the Roman Empire because he would have had to include dragons to the nth power. But you cannot have any theories with references to dragons under a closed system.

A Specialized Society

I find our society since World War I has specialized more and more. There has been an enormous amount of specialization, and we find our society assuming today that specialization is logical, desirable, and "natural." You learn that there are no alternatives. It is naturally that way—that is the way it is.

But I find life is born comprehensive—prone to be comprehensive. It seems perfectly clear to me that nature is so competent at designing specialists that if she had wanted man to be a specialist, she would have had him born with a microscope at one eye and a telescope at the other. What I find unique about man is not what he does with respect to any of the other living species—that is, physically with his internal organics—that is even mildly impressive. But where he is utterly unique is the way he employs the mind, and his development of the awareness of the extraordinary generalized principles that are operative in the universe, and his employment of them in special cases.

So man is able to discover, as Bernoulli did, the principle of pressure differentials, for example. What is unique, then, is the discovery of the principles, the employment of them, and the ability to exchange tools.

Man and Toolmaking

I did a case study of my hands. I can do things with my hands. I can cup my hands, but I need my hands for something else besides water. I found I needed water, all right, but when I went after berries I got very far away from water, and I kept getting thirsty. So I invented a vessel, and I can close it and I can carry it. This vessel can handle heats my hands cannot handle; it can handle acids my hands cannot handle; and I can make it a thousand times bigger than my hands—I can make it ten thousand times bigger than my hands. It begins to lose its similarity to hands and people lose the realization that this exists in the universe only by virtue of man. It's part of man.

Man has learned, then, how to externalize his own functions and to leave them behind. So that now you can use my hands, and we can go on from generation to generation of our hands, interchangeable hands. There are no tools that man has developed that are not extensions of the original integral functions, though the functions become, like the special cases in generalization, not too visible. They are always that way.

I don't find anything that has been done by man, that we call mechanics, that isn't part of his internal organism. He was apparently designed with this capability to externalize his internal metabolic regenerating organisms. And he is developing external metabolic regenerating organisms to take care of more and more human beings and extend the capability to all men so that all men can enjoy total resources no matter where they are.

There is something very big going on, and there is some-

129

t evolution is confronting man with that he doesn't
too well. I find very unsympathetic and short-
ements being made about technology and thinking-
as something very independent of man. It is not
many living species that develop external
instance, the bird's nest and the spider's web.
creatures, as with any system including inanimate systems, give off their energies and therefore alter the environment by doing so. Living creatures alter the environment a
little more as they give off more energies; they alter it much
more than the inanimates. The altered environment requires
alteration of the patterning of the living creatures. There is
the "epigenetic landscape," the interplay of living creatures
altering an environment and of an environment altering the
creatures. This goes on and on and it is what we mean by
evolution. It is inexorable and irreversible. Many creatures
alter the environment in nondiscrete ways. Other creatures
alter it in discrete and preferred patterns; as, for instance, the
bird's nest.

Toolmaking is the externalization in discrete ways aiding
the evolutionary process and the regeneration of the species.

Man is not unique as a toolmaker at all, but he is unique
in the degree to which this capacity is extended by virtue of
his mind and his ability to understand those generalized principles. And he is the only one to really alter those tools,
change those tools, and try to get better tools.

The Phenomenon of Technology

I am very eager to have humanists participate with me in my
feelings about the phenomenon of technology—a word that
is bandied about constantly and often thought of as the cause
of our troubles and pain. I do not see technology as something that is foreign to man. I hear the word "natural" and I
hear the word "artificial" and I am convinced that those
words are words of ignorance.

I am convinced that whatever nature permits is natural,
and that which nature does not permit, you cannot do. And
if nature has this as a generalized principle, it has in it the
option that man can employ to alter the environment to the
advantage of his fellow man. There are ways in which you
can alter an environment to decrease the freedoms of your
fellow man. But you can also go very far in increasing his
degrees of freedom and accelerating the rate at which he can
comprehend, communicate, and be effective. That is what we
are doing.

Craft and Industrial Tools

I have divided all the tools produced by men into two main categories, and they have helped me a great deal in differentiating out factors and problems. I call one craft tools and the other industrial tools. By craft tools I refer to all the tools that can be produced by one man starting nakedly in the wilderness without any information or aid from anybody else. So the stone becomes a tool; the stick becomes a tool. Then man makes a spear and it is even more effective—and he keeps modifying. These things the individual can develop out of his own personal experience, and he is prone to do so out of his own personal experience.

By industrial tools I mean all the tools that cannot be produced by one man. And I discovered a very important thing—that is, that the first industrial tool was the spoken word. So I think the spoken word was the beginning of industrialization, the beginning of the ability to relay one's experience to another man, to the next generation, when man began to compound his advantage. With craft tools, you have a very limited man, limited to where his own feet will take him, limited by his unevenly distributed resources. He is very limited in total experience and in time and capability. The industrial is quite the other way; it represents the integrated information of all men and all time. It's a very extraordinary power.

The Leonardo-type Man

There have been men in our history who have become well-known to us by virtue of their tool-inventing and tool-using capabilities, and their conceptualizing of tasks they could do—tasks they have done on behalf of their fellow men. I will simply call this kind of man the Leonardo-type. He was a very comprehensive toolmaker, tool-conceiver, tool-user, and a large problem addresser and solver.

In going back to our earliest known history of man we find that life was formidably difficult. Man knew so little. He had inbuilt hunger and thirst so that he would be sure to have the drive to get food, so that he would regenerate and reproduce himself. Man had little knowledge about what would support life. He didn't know what berries were poisonous, etc. The people who were relatively the strongest were able to overpower the animals. Men were beset with diseases. There was a great deal of fighting and struggling, and human beings died very young. By and large men could not rationalize the experience that life was meant to be an end in itself. Men, therefore, thought about the afterlife.

131

In ancient Egypt it was the afterlife of the Pharoah that absorbed so much energy. We find that Leonardo-types begin to emerge. You can see him in Egypt building pyramids for the afterlife of the Pharoah, his patron. Here the "scaffolding principle" developed; i.e., the tools invented and used in one period were retained for use by later builders. In ancient Rome great effort was made to take care of the afterlives of the nobles, later of the middle class, until finally the idea emerged during the Christian era that toolmaking should be used to take care of the afterlife of everybody.

In time the tool capability increased even more. They might not only take care of everybody, but take care of the living life of the King. This is really where the divine right of kings comes in—the new patronage of the Leonardo-type. This proliferated and in Magna Carta days people said, "We'll take care of the living life of all the nobles, too." Then this was extended to the middle class. This brings us right up to our present century and the point where the proliferation of tools is so great that the thinking man might be able to take care of the living needs of all men. Until our time the artist was making end products for the patron. It is only in our era that the process has changed, and that we produce not only for the patron but for all mankind.

In the middle of the 21st century Henry Ford may be identified as the great Leonardo-type, even though he would be absolutely astonished. He thought of his work as utterly prosaic, but he might then be thought of as having had Leonardo-type conceptualization. And the idea was that from this point on the artist makes tools and the tools make the end product, and this is mass production. That's where we've come to in our age. Since 1900 we've gone from less than one per cent of humanity to more than 40 per cent of humanity enjoying a higher standard of living than was known to any human beings—or dreamt of by any human beings—before the turn of the century. From less than one per cent to more than 40 per cent, and every bit of that has come indirectly as a fallout from technology.

The Building World and Performance Per Pound

I had been learning in the Navy as a comprehensivist what I found my fellow man did not seem to be aware of as a specialist. But I could see very readily as a comprehensivist in contradistinction to the land the world of building.

On the land we find men identifying security with heavier, wider, and higher walls. The psychology of man on the land is: the bigger, the more secure. On the sea, however, I found a completely different story. In building a ship you had to do things in terms of its floatability. And there is basic displacement, and you can only have a ship, whatever size of the ship and whatever that volume is, that's all the weight you can have in your ship and your cargo.

The great secret of the Navy through all the ages was never published. Nothing has been more classified than the information of what your ship did for the same amount of weight, for the same amount of time, same amount of energy, and same amount of muscle: How you could up your advantage. *They were continually doing more with less.* And now in the air, even more with less. And the more with less in the air in the last 60 years has been fantastic.

Now everything at sea was done in terms of performance per pound, or energy, or time. When I went into the building world, I found something that was extraordinary. I've been asked to speak to the architectural societies in almost every country of the world, and, every time I meet with architects, I will always ask: "Will you please tell me what the building we are in weighs? Could you tell me roughly within 100,000 tons? Tell me within a million tons?" I will say this to architects and get no responses. Quite clearly, if you don't know what a building weighs, you have certainly never been thinking about performance-per-pound.

I found, here on the land, man was thinking and operating out of fear and producing greatness and massiveness. The epitome of this was the Maginot Line, and the Maginot Line was suddenly and absolutely finished—whoever had the hardware, that was all. Fortresses have no meaning any more, but society is still thinking fortress, still thinking bigness. Performance-per-pound came from the sea technology and not from the land.

But if we are going to take care of everybody in the world with our extraordinary production capabilities, we are going to have to know something about our performance designs, and we are going to have to do more with less. I find that society not only has no book about this but that there is no chapter, there is no paragraph, there is not a sentence in any book in economics about doing more with less because it's the most highly classified idea that man has ever had.

This began to hit me very hard in my early days and by 1927 I said it could be that this concept of the Navy is a specialized case of a generalized principle of how to solve problems, which, if properly employed, might prove wrong Malthus's negative idea of man multiplying faster than his resources.

I'm perfectly confident that there is an ability to do more with less that makes it possible to do so much with so little, that we could take in everybody with a higher standard than anybody has ever known. There are aspects about technology that I feel are very, very important for humanists to understand. I think the universe has in it, waiting, the capability for man to become a success instead of, as Malthus assumed, a failure. One reason everybody loves babies is that when they are born they are so clearly designed to be successes, and everybody feels that they would like to have that chance at success again before they get all messed up. And we have that chance now. But it is going to be a design revolution, not a political revolution.

It seems perfectly clear to me that all of our society is operating in really great ignorance, and lacking understanding. I make a distinction between two classes of goods: what I call *weaponry* and *livingry*. By *livingry*, I mean that we use this great capability to actually make life a success. Yet we keep saying in great ignorance, we can't afford to do it. We don't care about the things that need to be done and we develop our technology only on the edges of war.

So part of the doing more with less would be to really produce the most extraordinary kind of matter control. I have done this with the geodesic dome, which is really an experiment of how you can do more with less. There are almost 10,000 of these domes in more than 50 countries in the world, many of them delivered by air. They are designed for full Arctic hurricanes and full Arctic snowloads and earthquakes. They have been standard-tested for any of those stresses, and they are good for all of them. And they weigh an average of less than three per cent of the weight per closed cubic foot of any known alternative engineering. They are getting even lighter too. Our whole building world, as we know it, is really dying. Our building world is the house—the shell costs more and more and is getting smaller and smaller, and it gets more expensive—expensive to the point where it is fantastic.

Concluding Remarks

I feel that evolution is intent to try to make man a success, and I find man in very great ignorance and not really understanding what is happening to him. Evolution is apparently intent, and the universe has made it possible for us to increase our performance-per-pound of the design revolution, to make it possible to enjoy the whole earth without interfering with another man or profiting at another's expense.

I'm quite confident now that we are going to have to be really on our own. I would be very worried about the whole thing if it were not for biology and anthropology and the fact that all human tribes and all biological species become extinct through over-specialization. Specialization is inbred at the expense of general adaptability. When you've lost adaptability, then you are extinct. Man was becoming more and more specialized and developing enormous capability to produce energy, with nobody to coordinate him.

We were becoming so specialized that we were about to lose, when suddenly one of our civilization tools, the computer, which is an extension of our brain and can operate faster than our brain, came into being. I'm quite confident that the great antibody to our specialization is the computer. The computer and what we call automation is about to take over the specialization and force man back to his innate comprehensivist role—to be really the humanist.

Women and work (III): The effects of technological change

by Madeleine Guilbert

Have the past fifty years of industrialization helped much to change woman's condition? Yes, in general. No, when it comes to employment. For, while industrialization has provided numerous new job openings for women, the work is still essentially of the same menial and poorly paid calibre as they have done since long before the spread of technology.

Will the new technology of automation change this picture? Not much, if present evidences can be accepted. Thus far, the introduction of automation in a plant has tended to wipe out low-skill jobs filled by women, creating either unemployment or else displacing the holders to other jobs still requiring the poorly paid feminine qualities of patience and manual dexterity.

INDUSTRIALIZATION AND EMPLOYMENT

Women have been taking employment outside the home since long before the development of modern labour-saving home appliances, which presumably freed them so they could do so. In France, women have been engaged in outside work since the Middle Ages; there were even guilds for women workers—kept under fairly strict supervision, it is true. In the eighteenth century, however, the number of working women increased considerably as the textile industry spread over the countryside and provided employment for peasant families in their own homes. Subsequently, as major industries came into being, from the first half of the nineteenth century onwards, ever-increasing numbers of workers began to flock to their factories—not only men, but women and children too, already conditioned by the textile industry and prepared to accept extremely low wages. The same sort of thing had occurred in England a century earlier.

Reprinted by permission of Unesco from IMPACT OF SCIENCE ON SOCIETY, XX:1 (1970).
© Unesco

By 1866, 34 per cent of the non-agricultural working population in France were women (2,775,000). This movement continued and, between 1866 and 1906, the number of women in non-agricultural occupations grew by 1,600,000, almost one million of whom were in industry. Thereafter, the number of working women remained more or less stable, though there were considerable modifications in the way in which they were distributed. In the traditional branches of production—the textile or clothing industries, for example—the number of working women decreased, as did the number of men. On the other hand, women played a greater part, both in actual numbers and in percentage, in growing industries such as the metal and chemical industries, where advances in mechanization facilitated their employment. At the same time there was a marked increase in the number and percentage of women in the tertiary sector,[1] an increase directly related to the progress made in the education of women. From then on, women who went out to work came from all social spheres; the diversification of the work available opened it up to new layers of society.

The changes brought about by the growth of industrialization were accompanied by deep-seated psychological alterations. The industrial era transformed people's views about work, and this was bound to lead to such changes. Work came to be regarded as an activity proper to the nature of man, something necessary for the harmonious development of his personality. Nowadays it is an exception —indeed almost scandalous—for a man not to work. Admittedly, this is not true of women; if they do not work outside the home, it is accepted that it is because they are doing housekeeping.

Nevertheless, employment outside the home has now become an important part of life for many women, even when combined with home duties. Indeed, their participation in working life is at the root of the problems relating to their participation in social and political life. In any case, it can be said that the development of industry, by making it possible for women to enter an increasing range of employment, has opened up to them ever-widening spheres of social activity and has brought about important changes in their status. At the same time, certain changes have taken place in family life, where women are increasingly taking over responsibilities which were once considered exclusively masculine.

It should also be noted that the development of communications technology —which in its turn is related to industrial development—has brought into the home the means of extending the range of women's interests. Women, who used to be almost completely ignored, have now become a very important audience for the mass communication media, particularly the press.

OUTSIDE THE HOME,
THE SECONDARY SEX

However, the changes in the status of women that have accompanied the devel-

---

1. The three sectors of the economy are: *primary*: agriculture and mining; *secondary*: industry and manufacturing; *tertiary*: services.—Ed.

opment of industry are far from being all for the good. All industrial societies are faced with problems concerning the status of women which are nothing other than signs of resistance to these changes. These problems take different forms in different types of society.

In some countries, anti-female discrimination is reflected in their institutions and in political, legal or professional matters. In others, while discrimination is not always apparent, yet there are many anomalies. Many examples could be given: inequality of access to education; inequality of access to vocational training; inequality in promotion; less pay for equal work; and less participation in political and trade-union activities. Even in the Socialist countries, which hold as a principle that there should be no discrimination against women, there are discrepancies in the allocation of political responsibilities, of work based on qualifications, and of professional responsibilities.

What are the elements involved in this state of affairs? First of all, there is the weight of tradition. It is a fact that, in all industrial societies, many women do not work outside the home, even in circumstances where economic necessity would seem to demand it. Some of these prefer to remain at home, even if their family obligations no longer take all their time. Others are more or less constrained to do so by social or family pressure.

The weight of tradition—the belief that most of the housework should be done by women—is still felt, even if they go out to work. A recent time-budget survey of families where women have outside jobs (Table 1) showed that on working days and even more so on Sundays women

TABLE 1. Average hours daily allotted to housework and child care (one or two children)[1]

	Wednesday	Sunday
Women		
Factory workers	3.87	7.12
Office workers	3.34	6.87
Professional workers	2.92	6.62
AVERAGE	3.37	6.87
Men		
Husbands of factory workers	0.38	1.37
Husbands of office workers	0.92	1.81
Husbands of professional workers	0.43	1.46
AVERAGE	0.49	1.65

1. M. Guilbert, N. Lowit, and J. Creusen, 'Enquête Comparative de Budgets-Temps', *Revue Française de Sociologie*, October-December 1965, p. 487-512.

still spend much more time on domestic work than do their husbands.

The effects of the evolution of industry in the last fifty years upon women's work must also be carefully considered. The growth previously mentioned in the number and percentage of women working in certain branches of industry is related to the move towards mass production and the more efficient use of labour after the First World War. This move, which began before 1914 in the United States, spread to other industrial countries; it led to an increase in the number of unskilled jobs as various tasks were redistributed and simplified. Many such jobs—the most simple, piecemeal and monotonous ones to which no responsibility was attached—were given to women.

In present-day France, the latest statistics available (April 1964) clearly show that the proportion of skilled workers is much higher for men (44.2 per cent) than for women (12 per cent). More women than men are semi-skilled workers, 48.7 per cent of them falling into this classification as compared to 32.8 per cent of the men. Only 1.1 per cent of women hold positions of responsibility and 1.8 per cent are technicians or supervisors, whereas 8.6 per cent of men hold positions of responsibility and 10.9 per cent are technicians or supervisors.[1]

In addition to such discrepancies in employment there are differences in vocational training. Amongst the industrialized countries, the most pronounced discrepancies are to be found in those countries which first underwent industrialization. In the Soviet Union, where statistics about vocational training make no distinction between young men and young women, it would seem that such training is in almost all cases provided for both sexes. In France, public establishments for vocational training do not, in theory, discriminate against women. In fact, although a circular has been sent from the Ministry of Education reaffirming this principle, young women generally continue to be refused admission to establishments preparing young people for types of employment regarded as a masculine, though lack of space and equipment is usually given as the reason. As a result, most vocational training for women in France is in office work and dressmaking, despite the fact that the number of women working in the dressmaking trade is steadily decreasing.

The situation is somewhat better at the professional level, but problems arise when it comes to finding employment. Women holding the same professional qualifications as men are rarely offered posts of like standing. Women are, for the most part, given less responsible jobs, which means different relations with other personnel. However, it is in the employment of women in senior posts in industry that differences between countries seem the greatest.

Such differences in the work done by men and women are often paralleled by differences in salary. This is so in France, Italy and Belgium, as well as in other industrialized countries. It is true that the discrepancies which were so marked at the end of the last century have diminished. Certain international conventions (for instance, Convention No. 100 adopted in 1951 by the International Labour Conference and Article 119 of the Treaty of Rome, which set up the European Economic Community) have laid down the principle of equal pay for equal work. But there are still discrepancies, especially in some of the Common Market countries. In France, the quarterly survey of the Ministry of Labour for 1 April 1969 showed that on the average women are paid 7.4 per cent less per hour, relative to the wages of men, for work requiring the same qualifications or of the same nature.

The industrial jobs that women generally obtain are those in which it pays to employ female labour, because of their special aptitude for simple, repetitive tasks, often acquired through their experience in domestic work. This discrepancy, as a

---

1. Figures taken from a quarterly survey of the Ministry of Labour, given in *Revue Française du Travail*, October-December 1966, p. 92-3.

result of which women are apt to specialize in certain types of work, obviously has profound economic implications.

The other side of the coin is that the weight of tradition has not been all-dominant. In some fields of employment, women are now obtaining jobs hitherto closed to them. The number of women in the liberal professions has increased. Women may compete for entry into certain of the specialized higher educational institutions, and there is a slight increase in the number of women holding senior posts or working as engineers or technicians, though there are very few comparable examples in the workshop.

In sum, therefore, the growth of technology, in so far as it has increased the part played by women in the working world and in society, has been a major factor in improving the status of women. This evolution still continues to meet with opposition, an opposition expressed with varying degrees of intensity, but perceptible in all industrial societies.

HOW WILL AUTOMATION AFFECT WOMEN?

We may now well ask: will the profound technological changes taking place around us—especially those related to the development of automation—be likely to speed up the changes taking place in the status of women?

It is at once obvious that technological changes promise increased mechanization and simplification of household work. Obviously, too, such changes could transform working conditions—shorter working hours, for instance, would be a logical consequence of the development of automation. Moreover, automation tends to change the nature of the work carried out. The possible effects of these three aspects of automation will be considered in turn.

The mechanization of housework in itself can hardly be said to have had a decisive influence upon the status of women in the past. As has already been pointed out, the entry of large numbers of women into employment took place before the technological developments which have simplified domestic chores. Moreover, the women who work are not always those who can afford to have efficient home appliances. In addition, the time-budget studies already quoted show that in many homes where women work they still do most of the housework. Therefore, technical means of simplifying housework, however important they may be, do not of themselves appear to be determining factors for changing the condition of women.

A substantial reduction in working hours as a result of the development of automation would probably affect the situation much more, and might help to overcome some of the difficulties which working women encounter. For instance, working women would have more time to undertake further training.

Now, will the changes brought about by automation in the nature of work itself enlarge the work possibilities for women and enable them to obtain posts presently generally refused them? This is the most important question. If we look closely at the present situation, however, we shall see that no definitive answer can be given. Automation is more developed in some sectors of industry than in others, and still affects only a few industries, even in the most industrialized countries. Furthermore, it is rare for an industrial firm to be fully automated; usually only one sector of a firm is automated, or else only some of the machinery installed is automatic or semi-automatic. Finally the industries using automated methods of production are rarely those where women workers are in the majority.

Why are industries employing a majority of women less automated than those employing a majority of men? Do the so-called female tasks in industry have certain characteristics which make them more difficult to automate? Or are there economic reasons—particularly the relatively low cost of female labour, which may minimize the savings to be achieved by automating a process—that help to brake the extension of automation to industrial sectors employing a large percentage of women? This may indeed be true of industry, but it does not apply to banking and insurance, where the work force is largely feminine and where computers are now in widespread use.

The effects of automation upon the working conditions of women have not been sufficiently investigated. The observations I have been able to make in this area concern one particular industry, the metal-working industry.[1] These observations showed that long-term effects on the work of women, which are very difficult to prognosticate, must be carefully distinguished from immediately observable effects which can, however, serve as an indicator. As for the latter, two distinct types of cases occur.

The first set of cases relate to work

hitherto carried out by women where women are still employed after the introduction of automation. Here, I observed the replacement of production lines for manual spray-gun painting by a chamber for automatic electrostatic painting and by an automatic conveyor belt for lacquering and printing flexible tubes. In these cases women were given the work of doing by hand what the automated process does not do, such as putting workpieces on the conveyor and removing them. The main requirements for such work are manual rapidity and precision and the ability to repeat exactly the same set of movements. This technological advance, therefore, has not really changed the nature of the labour done by women. Indeed, it has tended to emphasize its piecemeal, repetitive nature.

The second set of cases observed in the metal-working industry showed that when a process is automated female employees are partially or totally eliminated and replaced by male workers. When, for example, foundries installed sand-blowing machines to pack sand automatically into moulds to make cores for castings—a task which used to be done by hand by women —the result was a partial cut-down of female labour; the machines are operated by men, but women are still employed to remove the cores from the moulds, a delicate manual operation.

On the other hand, when automatic presses are installed, the women employees are gradually eliminated by stages. I observed this process in a bearing factory

1. M. Guilbert, *Les Fonctions des Femmes dans l'Industrie*, The Hague, Mouton, 1966, p. 222 ff.

being automated. In this particular factory, three types of machines were concurrently in use to polish bushings. Only women were employed to operate the oldest machines, in which the workpieces were fed in separately by hand. Both men and women were employed on the second type of machine, which was more modern and bulk-fed, but the women were there because they were old employees and only men were being taken on as new operators. The third type of machine, which was just being installed, was fully automatic; it needed no more than maintenance and supervision, and only men were employed for this.

Such examples appear to indicate that when automation is introduced the proportion of women workers is reduced. This is not to say that the installation of automatic equipment always leads to an over-all reduction of female employment. In fact, since automation particularly takes place in growing firms, women are often simply transferred to non-automated departments, especially those where rapidity and manual precision are still the main requirements. The same sort of thing occurs in office employment: a chain of stores, for instance, which has automated its accounting department, employs male programmers and operators for this work

and offers jobs as saleswomen to women hitherto employed there. It would seem, therefore, that in many cases the introduction of automation accentuates the tendency for women to be employed only on repetitive or unskilled work.

One cannot, however, be categorical about the possible effects of technological change, and the consequences at present observable should not be considered definitive. They are not irreversible, and it may be that the wider spread of automation will favourably affect the condition of women by opening up new fields of work for them. We cannot over-emphasize the importance of providing vocational training for women at a high standard and in a wide range of subjects, so that they can obtain the new jobs that are too often closed to them at present.

By pointing out the great changes which industrial development has brought about in the condition of women, by underlining the complex nature of the obstacles which still continue, in this era of science and technology, to block improvements in the status of women in industrial societies, I hope to have made it clear that many complex and far-reaching changes yet remain to be made.

THE CULTURE OF MACHINE LIVING

by Max Lerner

One could draw a gloomy picture of machine living in America and depict it as the Moloch swallowing the youth and resilience of American manhood. From Butler's *Erewhon* to Capek's *R.U.R.*, European thinkers have seized on the machine as the cancer of modern living. Some have even suggested that there is a daimon in Western man, and especially in the American that is driving him to the monstrous destruction of his instinctual life and indeed of his whole civilization.

Part of the confusion flows from the failure to distinguish at least three phases of the machine culture. One is *machine living* as such, the use of machinery in work and in leisure and in the constant accompaniments of the day. The second is cultural *standardization*, aside from the machine, but a standardization that flows from machine production. The third is *conformism* in thought, attitude, and action. All three are parts of the empire of the machine but at varying removes and with different degrees of danger for the human spirit.

The danger in machine living itself is chiefly the danger of man's arrogance in exulting over the seemingly easy triumphs over Nature which he calls

Reproduced from THE UNESCO COURIER, May 1971, pp. 23-27.

"progress," so that he cuts himself off increasingly from the organic processes of life itself.

Thus with the soil: the erosion of the American earth is not, as some seem to believe, the result of the mechanization of agriculture; a farmer can use science and farm technology to the full, and he need not exhaust or destroy his soil but can replenish it, as has been shown in the Tennessee Valley Authority (1), which is itself a triumph of technology.

But the machines have been accompanied by a greed for quick results and an irreverence for the soil which are responsible for destroying the balance between man and the environment. What is true of the soil is true of the household: the mechanized household appliances have not destroyed the home or undermined family life; rural electrification has made the farmer's wife less a drudge, and the mass production of suburban houses has given the white-collar family a better chance than it had for sun and living space. What threatens family life is not the "kitchen revolution" or the "housing revolution" but the restless malaise of the spirit, of which the machine is more product than creator.

EVEN in a society remarkable for its self-criticism, the major American writers have not succumbed to the temptation of making the machine into a Devil. Most of the novelists have amply expressed the frustrations of American life, and some (Dreiser, Dos Passos, Farrell and Algren come to mind) have mirrored in their style

the pulse beats of an urban mechanized civilization. But except for a few isolated works, like Elmer Rice's *Adding Machine* and Eugene O'Neill's *Dynamo*, the writers have refrained from the pathetic fallacy of ascribing the ills of the spirit to the diabolism of the machine.

The greatest American work on technology and its consequences— Lewis Mumford's massive four-volume work *The Renewal of Life*—makes the crucial distinction between what is due to the machine itself and what is due to the human institutions that guide it and determine its uses.

It is here, moving from machine living to cultural standardization, that the picture becomes bleaker. Henry Miller's phrase for its American form is "the air-conditioned nightmare." Someone with a satiric intent could do a withering take-off on the rituals of American standardization.

Most American babies (he might say) are born in standardized hospitals, with a standardized tag put around them to keep them from getting confused with other standardized products of the hospital. Many of them grow up either in uniform rows of tenements or of small-town or suburban houses. They are wheeled about in standard perambulators, shiny or shabby as may be, fed from standardized bottles with standardized nipples according to standardized formulas, and tied up with standardized diapers.

In childhood they are fed standardized breakfast foods out of standardized boxes with pictures of standardized heroes on them. They are sent to monotonously similar schoolhouses, where almost uniformly standardized teachers ladle out to them standardized information out of standardized textbooks. They pick up the routine wisdom of the streets in standard slang and learn the routine terms which constrict the range of their language within dishearteningly

_____
(1) A U.S. federal corporation created in 1933 to conserve and develop the resources of the Tennessee River Valley.

narrow limits.

They wear out standardized shoes playing standardized games, or as passive observers they follow through standardized newspaper accounts or standardized radio and TV programmes the highly ritualized antics of grown-up professionals playing the same games. They devour in millions of uniform pulp comic books the prowess of standardized supermen.

AS they grow older they dance to canned music from canned juke boxes, millions of them putting standard coins into standard slots to get standardized tunes sung by voices with standardized inflections of emotion. They date with standardized girls in standardized cars. They see automatons thrown on millions of the same movie and TV screens, watching stereotyped love scenes adapted from made-to-order stories in standardized magazines.

They spend the days of their years with monotonous regularity in factory, office, and shop, performing routinized operations at regular intervals. They take time out for standardized "coffee breaks" and later a quick standardized lunch, come home at night to eat processed or canned food, and read syndicated columns and comic strips.

Dressed in standardized clothes they attend standardized club meetings, church services, and socials. They have standardized fun at standardized big city conventions. They are drafted into standardized armies, and if they escape the death of mechanized warfare they die of highly uniform diseases, and to the accompaniment of routine platitudes they are buried in standardized graves and celebrated by standardized obituary notices.

Caricature? Yes, perhaps a crude one, but with a core of frightening validity in it. Every society has its routines and rituals, the primitive groups being sometimes more tyrannously restricted by convention than the industrial societies. The difference is that where the primitive is bound by the rituals of tradition and group life, the American is bound by the rituals of the machine, its products, and their distribution and consumption.

The role of the machine in this standardized living must be made clear. The machine mechanizes life, and since mass production is part of Big Technology, the machine also makes uniformity of life possible. But it does not compel such uniformity.

The American who shaves with an electric razor and his wife who buys a standardized "home permanent" for her hair do not thereby have to wear a uniformly vacuous expression through the day. A newspaper that uses the press association wire stories and prints from a highly mechanized set of presses does not thereby have to take the same view of the world that every other paper takes. A novelist who uses a typewriter instead of a quill pen does not have to turn out machine-made historical romances.

The answer is that some do and some don't. What the machine and the mass-produced commodities have done has been to make conformism easier. To buy and use what everyone else does, and live and think as everyone else does, becomes a short cut involving no need for one's own thinking. Those Americans have been captured by conformist living who have been capturable by it.

Cultural stereotypes are an inherent part of all group living, and they become sharper with mass living. There have always been unthinking people leading formless, atomized lives. What has happened in America is that the economics of mass production has put a premium on uniformity, so that America produces more units of more commodities (although sometimes of fewer models) than other cultures. American salesmanship has sought out every potential buyer of a product, so that standardization makes its way by the force of the distributive mechanism into every life.

Y ET for the person who has a personality pattern and style of his own, standardization need not mean anything more than a · set of conveniences which leave a larger margin of leisure and greater scope for creative living. "That we may be enamored by the negation brought by the machine," as Frank Lloyd Wright has put it, "may be inevitable for a time. But I like to imagine this novel negation to be only a platform underfoot to enable a greater splendour of life to be ours than any known to Greek or Roman, Goth or Moor. We should know a life beside which the life they knew would seem not only limited in scale and narrow in range but pale in richness of the colour of imagination and integrity of spirit."

Which is to say that technology is the shell of American life, but a shell that need not hamper or stultify the modes of living and thinking. The real dangers of the American mode of life are not in the machine or even in standardization as much as they are in conformism.

The dangers do not flow from the contrivances that men have fashioned to lighten their burdens, or from the material abundance which, if anything, should make a richer cultural life possible. They flow rather from the mimesis of the dominant and successful by the weak and mediocre, from

the intolerance of diversity, and from the fear of being thought different from one's fellows. This is the essence of conformism.

It would be hard to make the connexion between technology and conformism, unless one argues that men fashion their minds in the image of their surroundings, and that in a society of automatism, human beings themselves will become automatons. But this is simply not so. What relation there is between technology and conformism is far more subtle and less mystical. It is a double relation.

On the one hand, as Jefferson foresaw, the simpler society of small-scale manufacture did not involve concentration of power in a small group, was not vulnerable to breakdown, and did not need drastic governmental controls; a society of big-scale industry has shown that it does. In that sense the big machines carry with them an imperative toward the directed society, which in turn—whether in war or peace—encourages conformism.

On the second score, as De Tocqueville saw, a society in which there is no recognized elite group to serve as the arbiter of morals, thought, and style is bound to be a formless one in which the ordinary person seeks to heal his insecurity by attuning himself to the "tyranny of opinion"—to what others do and say and what they think of him. He is ruled by imitation and prestige rather than a sense of his own worth.

These are dangerous trends, but all of social living is dangerous. The notable fact is that in spite of its machines and standardization America has proved on balance less conformist than some other civilizations where the new technology has played less of a role.

Americans have, it is true, an idolatry of production and consumption as they have an idolatry of success.

But they have not idolized authority or submitted unquestioningly to human or supernatural oracles. They have had their cranks, eccentrics, and anarchists, and they still cling to individualism, even when it is being battered hard.

It will take them some time before they can become "man in equipoise", balancing what science and the machine can do as against the demands of the life processes. But where they have failed, the failure has been less that of the machines they have wrought than of the very human fears, greeds, and competitive drives that have accompanied the building of a powerful culture.

It has been suggested that the American, like the Faustian, made a bargain with the Big Technology: a bargain to transform his ways of life and thought in the image of the machine, in return for the range of power and riches the machine would bring within his reach. It is a fine allegory.

But truer than the Faustian bargain, with its connotations of the sale of one's soul to the Devil, is the image of Prometheus stealing fire from the gods in order to light a path of progress for men. The path is not yet clear, nor the meaning of progress, nor where it is leading: but the bold intent, the irreverence, and the secular daring have all become part of the American experience.

Women and technology in developing countries

by Barbara E. Ward

The tide of technological change—represented by labour-saving household devices, employment in factories, the conveniences of the towns into which people are migrating, increased education, and medical advances, particularly the Pill—is increasingly emancipating women in the developing countries from ancient ways of life which were characterized by heavy feminine burdens of toil. Yet, paradoxically, these instruments of liberation are at the same time causing many women to be chained more tightly than ever to their domestic duties.

At one level of argument the title of this article embodies a fallacy. Technological developments do not necessarily discriminate between the sexes any more than between colours, creeds or social classes. Changes in methods of transport and communication, increased control over epidemic diseases, availability of mass-produced goods, etc., are not intrinsically sexually selective. Already one cosmonaut has been a woman, and women are everywhere drivers of automobiles and aeroplanes, surgeons, precision workers of all kinds, computer operators and so on. Thus, even in its most dramatic aspects, the technological revolution is not an exclusively male preserve. We are all in it together.

Nevertheless, the prevailing division of labour between males and females inevitably brings it about that in most parts of the world there are certain aspects of technological change which, potentially at least, are of more direct significance for women than for men. Domestic organization is everywhere primarily a female responsibility. While changes in it certainly do affect men, too—and may indeed draw men into domestic chores more frequently than has been customary in the past—their effect upon women is always direct and often profound. Though less spectacular than flights to the moon or organ transplants, the developments in everyday matters (such as transport) and the

Reprinted by permission of Unesco from IMPACT OF SCIENCE ON SOCIETY, XX:1 (1970).
© Unesco

invention of new domestic appliances (such as refrigerators, rice-cookers, electric irons and the like—not to mention the Pill and other contraceptives) have had, or will have, a far more revolutionary impact upon domestic life and relationships and so, in the long run, upon the attitudes and values of both men and women almost everywhere.

THE EFFECTS OF TRANSPORTATION

A study of the changing roles of women in Asia, published by Unesco in 1963,[1] gives some vivid illustrations of the impacts of modern transportation on the lives of women in developing countries. A Thai woman gave a graphic description of the ways in which ordinary travel has changed since the end of the nineteenth century. A journey which took her grandfather two months on elephant-back and her father two weeks by train and on foot could now be accomplished in a few hours by fast car. An Indian writer in the same volume pointed out that modern means of transport have been one of the most liberating influences even upon women in full *purdah* (seclusion). And two women from countries as far apart as the Philippines and Ceylon explained in almost identical words how driving their own motorcars made it possible for each of them to combine successfully the roles of housewife, mother and professionnal educationist.

These writers were all highly educated representatives of their countries, and relatively well-to-do. However, the revolution in transport has affected less privi-leged women almost as much. For one thing it makes emigration possible, and is a kind of 'enabling clause' for urbanization. Where in the past the migration of workers, as in southern Africa for example, was mainly a man's affair, and women were left behind often for years at a time to carry on the home as best they could (with all the hard labour necessary for mere subsistence in a tribal society), the coming of relatively cheap and convenient mechanical transport makes the movement of whole families a fairly simple matter.

Recent migratory movements, like those from the Caribbean or the Indian sub-continent to the United Kingdom, though spear-headed by men, have very quickly included women and children as well. Here is a sphere in which 'modernization', which in its earlier stages was often entirely disruptive of family life, is proving far less so as it develops further.

At the same time, of course, more women than ever before are 'seeing the world'. It is still true that women travel away from their homes less frequently than men. Even today, my own village in south-western England contains several women who have never visited London and two who have never even been as far as the county town thirty miles away. None of the men has been so stationary. But these stay-at-home women are all over 60 years old. The younger ones have all travelled, some of them very far afield indeed. For good or ill they have made new personal contacts, seized new opportunities for education and employment, and have married men who in their grandmothers' days

1.   Barbara E. Ward, *Women in the New Asia*, Paris, Unesco, 1963.

would have been considered 'foreigners'. Similar changes are taking place all over the world.

THE SPREAD OF EDUCATION

But travel is only one part of the modern system of communications. Not only people, but goods and ideas too, are being distributed more and more freely. The spread of books, newspapers and the telephone, radio, cinema and television marks a whole series of social revolutions. In education, the arts and entertainment, their influence is obvious, producing new knowledge, new concepts, new and modified social attitudes, new ways of passing time and new openings for employment. The volume already referred to points out that the husbands who refused to countenance the education of women because their wives might learn to send and receive love letters from other men foresaw only a very restricted few of the complex transformations that improved communications would bring.

One of the most striking accompaniments of that side of technological change which is connected with the spread of modern education has frequently been a widening of the intellectual gap between women and men, and—even more damaging for those it affects—between parents, especially mothers, and their sons. At its most drastic this latter gap may involve an almost complete breakdown in communication. The tragedy implicit in such a situation probably never affected very many families, and will in any case affect fewer and fewer as the numbers of edu-

FIG. 1. As this policewoman of Lagos (Nigeria) illustrates, with the gradual emancipation of women in developing countries, jobs traditionally considered to be masculine are opening up to them.

cated girls increase, but there are still some women in Africa and Asia whose sons were sent to school in Europe so young that they find it almost impossible to talk with their mothers in their native tongue when they return home.

Although the generation gap is seldom as wide as this, nevertheless during the years in which modern schooling is just becoming universal it is inevitably very great. Moreover, because the education of girls lags everywhere behind that of boys, women are likely to remain the greater sufferers for at least a long time to come.

The educational discrimination between the sexes affects relationships within the same generation, too. Marriages which break down because of incompatibility in this sense are not few. Any discussion of modernization which ignores problems of this nature (and they are many) would be seriously incomplete.

Even though it continues to lag behind, however, the education of women has probably been by far the most important factor in bringing about changes in their role and status all over the world in the last fifty years. It is education, moreover, that helps make it possible for women to take advantage of the many technological developments that can provide them with better material goods and save them from hard and time-consuming physical labour. Women who cannot read or keep accounts cannot successfully cope with urban life, public transport and modern kinds of shopping; nor can they understand the printed instructions which accompany so much of modern merchandise. Literacy gives them access not only to books and newspapers and letter-writing, but also—and much more important—to bus and train services, modern shops, and a whole new range of goods and services.

THE PRODUCTS OF TECHNOLOGY

The goods and household equipment that are now available are contributing to other aspects of the domestic revolution. Cloth, which in many places and until very recently had to be hand-woven (even, as in much of South-East Asia, hand spun) can now be bought by the yard; sewing machines speed the making up into garments, and even provide a source of income; mass-produced enamel, aluminium and plastic pots and pans are to be found all over the world; soap, cosmetics, medicaments, surgical plaster, even comfortable and hygienic sanitary towels, are on sale almost everywhere. For people who live in towns foodstuffs, once all laboriously planted, weeded, harvested and processed by family hand labour—largely female—are now available, ready-wrapped in the stores. Even in villages the preparation of food has in many places been made much easier by the introduction of electricity and piped water.

These things do not by any means exist everywhere in the world as yet, but they are surprisingly widespread, and where they do exist their effect upon women's lives can hardly be exaggerated. For example, there is a fishing village in Hong Kong (typical of many others), which I first visited in 1950. At that time all the fishing-boats were wind-driven, slow and at the mercy of the very uncertain weather of the South China Sea. All the fisher families lived entirely on their boats, owning no property whatsoever ashore. The village contained no radio, no piped water, no electricity, no latrine. Only two of the fishermen's children went to school and both of them were boys.

By 1966, all but one of the boats had been mechanized, and each had a radio. Many fishing families had moved their older members, most of their women and all their younger children into newly built houses on the land. There was a new school, with three trained teachers and more than a hundred fisher-pupils, including all the girls between the ages of 7 and 15. There were also a splendid latrine and three stand-pipes with a never-failing supply of sweet water. In addition, there was a new maternity and child welfare clinic in the nearby market town, at which all the babies for the last five years had been born (previously all births had taken place on the boats, at sea).

In 1968 the generosity of a charitable trust in New Zealand made it possible for the villagers to purchase a small but adequate generator which gave them their electric light. Within six months every house had an electric iron, an automatic rice-cooker, and at least two electric fans to set beside its radio. One enterprising shopkeeper was selling iced drinks from his electric refrigerator, and there was talk of installing television.

In less than twenty years the women of this village have experienced a technological revolution which has improved their and their children's health, increased their expectation of life and given their children the chance of a good education. It has also improved their personal income, for now, for the first time, they have a good deal of leisure which most of them use to earn extra cash by making plastic flowers at home for a Hong Kong factory.

The women themselves say that the changes they value most are those which have brought better health and the chance of an education for their children. They also mention their appreciation of having more leisure and less everlasting hard work. They are quite well aware of the advantages technological change has brought to their own lives, discuss it clear-mindedly and often, and are busily looking forward to opportunities of buying still more electrical equipment.

On the other side of the world, in Mexico, changes almost as far-reaching have taken place in the same period. The key technological innovation in the villages there was a very simple one: the introduction, about fifteen years ago, of a mechanical method of grinding maize.

Maize, as used for making the flat, pancake-like *tortillas* which are the staple food, must first be soaked in water and then ground daily. Up till about 1950 most village women had to rise every day before dawn and trudge off to the natural rocks which were their grinding places. Grinding was done by rolling the maize between an upper stone held in the hands and the living rock. For an average household this took two or three hours—sometimes longer—and it was tiring work. Today the housewife simply takes her bucket of soaked maize along to the local mill, usually situated in a small shop. There she joins her neighbours in a friendly gossip until her turn comes round, and then the job is done for her—quickly, cleanly and cheaply.

The housewife also has piped water in her village now. She probably owns a sewing machine and—like her counterparts in the Hong Kong village—an electric iron, a radio and often a television set. She is still a very busy person and many aspects of the material conditions of her life would appear intolerable to a contemporary villager in a fully industrialized country. Nevertheless, her life is already vastly different from her mother's. Her daughters will undoubtedly see further changes, too.

152

The examples of change cited above—which stand for many—involve villages. However, it is obvious that the effects of technological change are still more apparent in towns, especially in the great metropolitan cities which have been developing at an unprecedented rate on every continent during the present century. It is here that the full impact of modern technology upon everyday living is experienced. Here are the centres of manufacture and distribution of material goods, and here are such large concentrations of population as to make at least some supply of services essential.

The surge to the towns, which is one of the major characteristics of our time, is to be explained as much by the 'pull' exerted by all these attractions as by the 'push' from an overpopulated countryside, for even where the countryside is not overpopulated (in Thailand, for example, or many parts of Africa) the towns continue to attract an apparently endless stream of rural emigrants. The reasons for this are made clear in the following true cases.

Two girls from the Hong Kong fishing village described above have married townsmen. One of them lives with her husband and three children in a small one-room apartment, measuring about 12 feet square, in a recently built block on an estate containing several thousand identical apartments. The family shares communal lavatory and washing facilities, but does all its cooking, ironing, mating, homework, sewing and sheer living in the one room, surrounded in the building by others with whom the family had no previous acquaintance and enveloped always by a volume of noise that has to be heard to be believed.

This woman has asserted over and over again, both spontaneously and in answer to direct questioning, how happy she is to have escaped from village life. Here, she says, everything is convenient. She and her husband own a refrigerator, two rice-cookers, an electric iron, a small electric heater (Hong Kong can be cold in the winter) and a television set. The water taps and flush toilets, though shared with others, are only a few yards away. A street market, shops and a small restaurant are at her door. The apartment block also houses a primary school, a clinic, a welfare officer and a social club. The rent is very cheap.

The other girl is less lucky. She lives in a squatter's hut half-way up one of Hong Kong's steep hillsides. Made of wood and beaten-out kerosene tins, the hut is far from weatherproof, and in the rains the earthen floor becomes a pool of mud. There is no toilet. The nearest stand-pipe for water is half a mile away, and except in the dry season it is easier to fetch water from a near-by (very doubtfully clean) stream. The three children are well dressed but dirty. So far she has managed to find a school place for only one of them. Her husband is intermittently unemployed.

Despite all this, this woman, too, frequently declares her firm conviction that she is far better off here than in the village. Pressed for a reason, she points to her electricity supply (probably illegally installed), her two rice-cookers, electric iron, fan and heater, and the sewing machine with which she makes a little extra money. On being informed that she could also have all these amenities in the village now-a-days, she explains that life here in town is far less hard work than it ever was in the village, that town shopping is convenient, public transport useful, and the local welfare clinic on the spot. Finally she adds that she has far more leisure in town.

These two case histories are included in order to draw attention to the fact that modern technological innovations make it possible even for people living in what by most standards would be considered insupportable conditions of squalor and overcrowding to consider themselves much better off than they would have been in traditional circumstances. That there are enormous numbers living in towns (and villages) today who do not have access to the goods which would make them feel like this is true. The brutalities of acute poverty and actual hunger are still real enough. None the less, it remains true that more people, in absolute terms, have more goods than ever before, and that many of these goods liberate women in particular from at least some of the back-breaking work that has been their usual lot, and that more of the people so affected live in urban than in rural areas. Women to whom goods of this kind mean not just prestige but also less fatigue, better health and greater leisure are usually quite clear in their minds on this point. Generally speaking they want to live in town.

It is in towns, moreover, that women are more likely to find employment, though it seems likely that unlike men, who go to the towns to seek work, most of the migrant women who take employment in the cities of the world do so rather as a response to the economic necessity they find pressing upon them after they have arrived. And when wage-earning in town takes places, its effects are far-reaching, for it is likely to be in a non-traditional occupation, will probably be connected with modern technology and will bring with it far-reaching changes in domestic roles and in outlooks. Women working in modern textile mills, for example, or in a big store, business office or educational establishment, have to keep to fixed hours and their pay is usually based on hours worked. Because they have prior obligations to housework, catering, cooking and, above all, children, most women find this both difficult and tiring.

It is sometimes argued that the organizational problems that outside work poses for women are more easily dealt with in countries where the large, extended family exists, or where domestic servants are still easily available. This is probably true, but it does little to help the majority of working women who in any country come from the poorer sections of the population among whom both extended families and domestic servants are rare or non-existent.

Even among the upper classes in such countries the problems are becoming fairly large. With more and more varied employment available and wages rising, the attractions of factory work prevail easily over the few perquisites of domestic service. At the same time, the move towards 'going separate', that is setting up independent simple (nuclear) family households is very marked.

I have heard highly educated Asian women debating in worried tones the

154

same problem that worries their Western counterparts: now that at last we are educated and emancipated we find ourselves chained more securely than before by the sheer demands of domesticity. For such women one can predict an increasing resort to modern domestic gadgets.

The technological revolution, with its educational and other 'modernizing' corollaries, first frees some women from total immersion in the domestic sphere, then in freeing others begins to confine the first group again. Finally by producing still more technological equipment of a domestic nature, it may help to redress the balance it has itself destroyed.

THE NEGATIVE ASPECTS OF TECHNOLOGICAL CHANGE

Sometimes the impact of technological change has been unmitigatedly depressing to standards of living. This was very frequently the case in nineteenth-century Europe, for example. It was the basis of Dickens's attacks upon English society, and of much of the work of Engels and Marx, to mention only three of the most influential writers.

Relatively secure in their Western welfare states, with their remarkable new washing machines, detergents and deep freezes, bourgeois Western writers try to believe that such things cannot happen any longer. The world is supposed to have learnt its lesson; industrialization is expected to proceed in an orderly fashion, taking due care of the human relationships and the human wants of the workers. Yet in every continent there are back streets and shanty towns which, despite everything said above about the ways in which they can be made supportable, are a disgrace to civilization. A very large proportion of the hundreds of thousands who flock into towns live there below the poverty level. How can one decide whether or not they would have been better off if they had stayed in their villages? Put very simply, industrialization and urbanization still can, and very frequently do, depress as well as elevate standards of living.

Modern technological developments can have their Luddite effects too. Cottage industries (in which women are able to take a full part mainly because they involve work at home) are especially vulnerable. The story of the nineteenth century English hand-loom weavers is well-known: how, under their possibly imaginary 'General Ludd', they fought against their displacement by power looms, resulting in widespread unemployment, lower wages and shoddier goods. Exactly similar reactions have occurred all over the Orient.

It does not take the full force of large-scale foreign competition to drive small peasant industry out of business. The impact of local technological change can be just as dramatic, as the following illustrates.

Until about fifteen years ago the Melanau women of Sarawak, in East Malaysia, enjoyed a somewhat unusual economic and social independence because their individual contribution to the processing of the Melanau's single export crop, sago, brought them in a regular cash income. Then, about 1955, an enterprising Chinese invented a mechanical means of refining sago. Within a very few years the need for women's employment had almost entirely disappeared—and there was nothing to take its place. The Melanau women gained enforced leisure, and lost their incomes; relief from toil meant also loss of independence.

The facile optimism which sees nothing but good in the increasing pace of technological advance is as obviously ill-founded as the equally facile pessimism which sees nothing but bad. Their circumstances being so different, Melanau and Hong Kong village women would be hardly likely to see eye to eye on this matter.

MEDICAL ADVANCES AND BIRTH CONTROL

There is, however, one field where the impact of modern technology upon the lives of women appears, on the face of it, if not unmitigatedly good, at least very widely welcome. This is the field of health, where medical advances have brought many improvements. Probably the most important of all medical advances is birth control, which is undoubtedly one of the crucial factors to be considered in any assessment of the effects of modern technology upon women. For the first time in human history there is the possibility of freedom from the physiological and social effects of the more-or-less continuous round of pregnancy, parturition and lactation which, with numerous miscarriages and the ever-present chance of death in childbirth, has been the lot of the vast majority of women between the ages of 15 and 50 since the human race began.

Moreover, the babies that are born are likely to survive. This, too, is crucial. A modern Westerner reads the pathetic inscriptions on the numerous tiny graves of past centuries which are scattered throughout the old burial grounds of Europe and North America with pity; most Asian, African and South American women would read them with a sympathy born of experience. One of the most striking differences between conversations with women in most of the so-called developed countries and most of the so-called developing countries today is that in the former one can ask: 'And how many children have you had?' whereas in the latter you ask: 'How many have you raised?' Increasingly as the former question becomes safer to ask in the developing countries—and this is happening—the roles of wives and mothers there will be marked by a new freedom from fear.

These are matters of peculiar personal concern to women. Their impact upon the structure of domestic relationships and thence upon the whole fabric of society itself is barely to be seen as yet. All that can be said at this stage is that the effects when they do become apparent are likely to be very considerable, unprecedented and, because of differing domestic circumstances in different parts of the world, various. And they will not be confined to women.

Schools are going metric

*A retired research chemist, Fred Helgren has been an officer in
the Metric Association, an organization that prepares and distributes
metric educational aids. He has been active in promoting the adoption
of the metric system to replace the several systems now in
use in the United States.*

Forget the length of King Edgar's foot, the length from the nose to the tip of the finger, the length of three barley corns laid end to end, the amount of land that can be plowed by a yoke of oxen in one day. Forget, if you have not already done so, the number of square feet in an acre, the difference between a dry quart and a liquid quart, the number of pecks in a bushel, and all the rest of the system of measures that are learned with difficulty and forgotten with the greatest of ease.

The legal system of measure in the United States is actually the metric system. It was adopted by an act of Congress in 1866. Children should have been educated in this language of measure following that important step in the improvement of our systems of measure. Charles Sumner, the senator from Massachusetts who sponsored the Metric Bill of 1866 that made the metric system legal in the United States, stated at that time, "They who have already passed a certain period of life may not adopt it, but the rising generation will embrace it and ever afterward number it among the choicest possessions of an advanced civilization."

The sciences saw the advantages of the metric system, adopted it, and have used it almost exclusively. It was not, however, accepted for general use, and schools approached the use of the metric system in a way that gave it little encouragement. The following are some of the poorly conceived practices:

1. Metric measure was not studied as a system by itself.

2. People were not taught to THINK METRIC.

3. Textbooks often contained only a single unit on the system, and problems were merely conversions from one system to the other.

4. The unit on the metric system was frequently at the end of the textbook. As a result, it was seldom taught. Teachers had little knowldege of the system, and it was omitted because of lack of time.

Now a new day has dawned. Following the three-year metric study by the National Bureau of Standards and the Secretary of Commerce, Maurice H. Stans, who was Secretary of Commerce in 1971 when the study was completed, made the following recommendations to the Congress of the United States:

- That the United States change to the International Metric System deliberately and carefully;
- That this be done through a coordinated national program;
- That the Congress establish a target date of ten years ahead;
- That there be a firm government commitment to this goal;

THE ARITHMETIC TEACHER, April 1973, 265-267.

- That early priority be given to educating every American schoolchild and the public at large to think in metric terms.

There is no need for our schools to wait for the Congress to act. There is good reason to feel that they will act, setting a target date ten years ahead, beginning in the year 1973. The metric conversion bill, S-2483, was passed in August 1972 by unanimous voice vote in the Senate. The House can be expected to vote for the change, and the President supports the bill.

What should be done in the field of education to GO METRIC?

1. Teach the metric system by itself so that teachers and pupils learn to think in this language of measure. Do not try to learn or teach the metric system through conversion problems, and do not try to learn conversion factors. Learn the metric system by itself. THINK METRIC.

2. Change mathematics and science textbooks so that only metric units of measure are used.

3. Before textbooks are changed, get metric workbooks for each teacher and each pupil. Then the system can be learned with very little individual effort.

4. Select one member of the faculty to be the metric authority for the school. He can get the information and materials necessary to enable the school to GO METRIC.

5. Encourage teachers to become members of an organization that will send them literature that explains the metric system, provides information on sources of educational aids, and publishes a newsletter that will keep them alert to metric progress and developments in the teaching of units of measure and their use.

6. Teach the metric system to all prospective teachers, for the change to the new system of measure is not just a mathematics or science project.

The working units of the metric system are easy to learn. The unit of length is the meter (or metre); the unit for mass is the gram; and the unit for volume, the liter (or litre). To THINK METRIC, it is well to learn the three basic units in combination with the prefixes *milli, centi,* and *kilo.*

$$milli \text{ means } \frac{1}{1000}$$

$$centi \text{ means } \frac{1}{100}$$

$$kilo \text{ means } 1000$$

For practical purposes, the whole system can be summarized as follows:

1000 millimeters (mm)	= 1 meter (m)
10 mm	= 1 centimeter (cm)
1000 m	= 1 kilometer (km)
1000 milligrams (mg)	= 1 gram (g)
1000 g	= 1 kilogram (kg)
1000 kg	= 1 metric ton (t)
1000 milliliters (ml)	= 1 liter (l)
1000 cubic centimeters (cm³)	
	= 1 cubic decimeter (dm³)

The milliliter and the cubic centimeter have the same volume. The term *kiloliter* is not recommended—it is equal to a cubic meter (m^3), which is more easily understood and used.

Machinists measure in millimeters; tradesmen measure in centimeters and meters; clothing sizes are given in centimeters. And greater lengths are in meters and kilometers. Mass is measured in milligrams, grams, and kilograms by the chemist; grams and kilograms by the shopper; and kilograms and metric tons when large quantities are involved. (Mass is the quantity of matter, whereas weight is a force, the earth's attraction for a given mass. Generally, the term *mass* is meant when we use *weight.*) Physicians, pharmacists, chemists, and bacteriologists use the terms milliliter, cubic centimeter, and liter. Consumers will make purchases of gasoline and other liquids in liters; large quantities will be sold by the cubic meter. Through everyday uses the metric system can be learned in a short time.

In conclusion, I repeat the recommendation made by the former Secretary of Commerce, Mr. Maurice H. Stans, that early priority be given to educating every American schoolchild and the public at large to think in metric terms.

For the price of $3, the following materials can be obtained from the Metric Association, 2004 Ash Street, Waukegan, IL 60085, a nonprofit organization interested in the dissemination of metric educational materials and information:

One 20 cm plastic ruler

Two 1.5 m plastic, flexible measuring tapes

Two copies of *Metric Units of Measure,* a booklet

One copy of *Metric Supplement to Science and Mathematics,* a workbook for use by the teacher and the pupil

One GO METRIC bumper sticker

One price list of metric educational aids

A copy of the last newsletter

An annual membership in the Metric Association that includes a subscription to the quarterly newsletter

Technology, Technique, and the Jesus Movement

The Jesus movement's often shrill denunciation of the churches issues from a conviction that we church people have sold our simple faith in Jesus for a mess of pottage.

W. FRED GRAHAM

✦ PROPONENTS of "counterculture" styles of thought and life differ critically regarding the place of technology in the new or "green" world that is nearing parturition. On the one side are those who — like Charles Reich (*The Greening of America* [Random, 1970]) or Jean-François Revel (*Without Marx or Jesus* [Doubleday, 1971]) — laud the products of technology and insist that the new society will use them without being dominated by them. Indeed, they hold that some of these products — the pill and the tape deck, for instance — were a major influence in creating the new consciousness and that they are necessary if Western man is to topple the Corporate State and come into new openness and freedom. On the other side are those who — like Theodore Roszak (*The Making of a Counter Culture* [Doubleday, 1969]), the compilers of the *Whole Earth Catalog* and admirers of Carlos Castaneda's three books on the Yaqui Indian *brujo* Don Juan — believe that technology is the enslaver of modern man and that, unless we shrug off its products as well as the rationalistic mind-set which gave rise to technological society, there is little hope for the future. Century readers will see the shadow of Jacques Ellul (*The Technological Society* and a spate of related books) behind this position.

In the more specifically religious field, these two attitudes toward technology and its manipulative corollary are much in evidence. Broadly speaking, it can be said that disciples of the Maharishi and researchers in brain-waves-*cum*-meditation represent the "pro" side of machinery and know-how, and the Jesus movement represents the polar opposite. Let us look first at the former.

I

In classic Eastern mysticism the aim is *moksha*, the existential realization that the essence of the individual (*atman* or soul) and the Essence of the universe (*brahman*) are one. The Buddhist term for this realization is Buddha-consciousness, though Buddhism holds to a greater philosophical skepticism concerning what it is that is united to the Holy Nothingness of Nirvana. The route in either case is long and difficult and somewhat painful. In Hinduism, such realization can be attained only by one who has studied long under a guru and has followed one or other of the arduous yogas or disciplines (the "yogas-yoke"). In Buddhism, the choice of yogas has been narrowed down to *raja* or "royal" yoga; i.e., the discipline of meditation.

But when the West really got into the Eastern thing, it was not through the long discipline of a yoga. Not at all surprisingly, the West achieved instant enlightenment by way of drugs. More recently, it has invented Transcendental Meditation as a way of getting quick results. This way looks genuinely Oriental, but whereas no Eastern religion has ever promised practitioners of its disciplines that kind of rapid success, Transcendental Meditation guarantees peace of mind to those who practice it for two 20-minute periods per day. Thus efficiency, the chief mark of technique, links up with Eastern religion the very moment it begins to have widespread effect in the West. In other words, it would appear that Eastern religion will not have even a ghost of a chance in the quick-production, quick-consumption West unless it streamlines itself to the point of caricaturing its historical and philosophical integrity.

How easily technique, with its god efficiency, can move into religion is illustrated by the current scientific study of altered and expanded consciousness. Here one runs into the tangled wires of devices that measure skin temperatures and brain waves. (For a summary of what is happening in this field see Bill D. Schul's "Altered Consciousness: What the

161

Research Points Toward," in the January 19, 1972, Century.) However interesting they may be, a caveat is in order regarding all such "scientific" studies. Remember, for instance, Masters and Johnson's research on sexual intercourse. It is probably true that because of their pioneering work people can be taught to have more satisfying sexual experiences; but it seems equally true that this may have little or nothing to do with loving another person. I am saying that we of today tend to substitute what technique *can* do for what it cannot do; we substitute sexual technique for love — and then wonder why love falls short and sexual techniques finally fail.

II

One of the interesting things that electroencephalographology has demonstrated is that the human brain puts out at least four kinds of electrical current, which researchers dub alpha, beta, delta and theta waves respectively. It is the theta waves that technicians are particularly excited about. They believe that these are the same waves that account for the visions of the great mystics, and they say that people who have been practicing meditation for a long time also produce theta waves. To be sure, these scientists find that alpha waves too indicate a turning-off of outside stimuli and concentration on one's own thoughts; but they think the former are more common and represent a level of consciousness lower than that indicated by theta waves. Anyway, serious research is now going on to discover what the connection between science and religion is in the human brain. One quotation may be sufficient to suggest the importance of this research for religious practice. Dr. Elmer Green of the Menninger Foundation says (in the article by Bill Schul):

> The electroencephalograph, along with other devices, for the first time has provided science the opportunity to examine human consciousness. We are learning a great deal about the human mind, how it functions, how it learns, and this, in time, should have quite an impact on the whole educational process. But perhaps of even greater importance, science is beginning to confirm many of our religious and mystical traditions. In the past science has been criticized as the antithesis of religion, but today a body of evidence is growing which indicates that the higher levels of conscious awareness are not only a reality but also a legitimate business of science.

A friendly statement indeed from the scientific community! But suppose you examine religious experience by way of the techniques described (plus "brainwave feedback training") with a view to enabling a person to attain in a few weeks a point of development in reaching higher modes of consciousness that (as Schul quotes researcher Erik Hoffman as saying) would have taken "three years under any other discipline." I wonder whether this is not to bend religious experience to the technique you are using. No doubt the techniques will work; no doubt people can be taught to put out theta waves at a rate not matched since Isaiah saw the Lord "high and lifted up." I can hear technicians exclaiming: "Man, look at those theta waves! It's a new record, and set not by some Zen master but by a white middle-class Jewish sophomore from MIT who trained hard two hours a day for six weeks." But just as I know from 20 years of marriage that love includes but is infinitely and almost undefinably more than sexual technique, so I suspect that anyone who is serious about man or God, about *moksha* or Buddha-consciousness, will not see in that wavy line any evidence for what he is passionately concerned about. Most people, however, *will* see that line as evidence. Because what technology and know-how cannot get you, and get you quickly, is not to be got; anyone who disagrees is simply afraid that the secrets he has worked for so long and hard can be appropriated by the masses in six easy lessons. Technology and propaganda of a well-meaning sort have programmed us to respond this way, for you can't sell the products of our industrial process by telling people they are hard to get! But so far in human history the result of technique has been not a heightening but rather a narrowing and flattening of human experience.

III

So we come to the antitechnology group of the counterculture — to the Jesus movement as representing the amorphous, many-faceted, anti-institutional and fundamentalist ferment among college-age youth today. The first thing to note is the soil in which this particular "greening" has grown. Many of our young people find it difficult to be as optimistic about life in our changing society as

Reich and Revel profess to be. Indeed the optimism of *The Greening of America* seems remote from the mood of introspection and pessimism that many of my students express in talks with me. As they see it, their own perceptions are systematically ignored by the world, and their only alternatives are to drop out — and thus give up any ideas of influencing the system; or to stay in — and thus run the risk of having their minds so warped that they won't want to influence the system.

But I think I detect an even deeper pessimism in my students: their feeling that they are living in a cold, meaningless universe in which man emerged from the haphazard evolutionary process by accident, and a sorry accident at that. Either the universe "means something" or it does not; and the suspicion of many is that it does not.

This, of course, is the starting point of nontheistic existentialism. The message of Jean-Paul Sartre is, in effect: "The only meaning you will find in an accidental universe is the one you impress upon it." But that is quite an effort when the end is not rebirth or life with God, but disappearance into the earth — "enriching the humus," as theologian Sam Keen says. Our destiny is the same as that of the billions of people who have come and gone on this planet since personality came into existence — most of whom never had leisure to look up from their toil, to become "self-actualized" and thereby find some fleeting meaning for themselves. With God dead, the sacred cosmos of the past rendered utterly profane, and once-great theologies, East or West, turned into caricatures, man is thrust back upon his own tiny resources for creating and holding onto meaning. Is it any wonder that sorrow rather than joy and happiness seems to characterize American youth? There is pathos in what is dead: "Bye, bye, Miss American pie."

The Jesus movement seems to be an answer to this emptiness — an answer by way of a narrowing and deinstitutionalizing of traditional Christian theology. And perhaps inevitably such. For on American soil the church never held a place comparable to that which it held in Europe. Moreover, traditional theology has always been simplified here. People who had a continent to conquer could spend little time shaping and elaborating a theology. Their question was: How much do you need to know to be

164

saved?

Modern youth's questions are less direct. One can imagine the following dialogue: Q.: What good are my efforts to change the world when nothing changes? A.: God will change it in his time, not ours. Q.: What good are theta waves if they only amount to psychic masturbation to the rhythm of a machine? A.: Jesus shows us that God is the Encountered One in every religious experience; but he needs no machines, only trust and prayer. Q.: What good are sensitivity groups if the world remains insensitive? A.: Be part of a movement whose aim is love, whose foundation is love, whose Exemplar is love. Q.: What about the hole in my life when all I can find to fill it with is ultimately empty too? A.: God promises you eternal wholeness. Q.: How can I live in a world, whether under capitalism or communism, whose methods deface the beautiful and crucify the good? A.: Come, you'll see that with God's power every crucifixion can become a resurrection.

IV

Writers about the Jesus movement are quick to point to its origins in biblical Christianity and to emphasize that the young are trying to live out what their parents professed to believe but did not practice. But the strong antitechnological bias of these college-age Christians is often missed. In their search for a simple Christian existence, they are like their non-Christian fellows in the counterculture who eschew as much of modern society as they possibly can. It is as if they had a sure instinct that today's unfaith is rooted largely in an industrial process; that — as Ellul has argued so persuasively and as most of us experience existentially — the technological society is radically incapable of faith.

How so? Well, we hear a great deal about agricultural and fishing communes which attempt to reroot people in nature and make them marginal to industrial — or, if you prefer, postindustrial — society. But it is more instructive just to observe and listen to these young men and women (they are neither shy nor dumb). They are not interested in jobs which take time from witnessing; for them, as for St. Paul, work is a means of holding body and soul together so that they can carry on the real purposes of life.

Their quarters lack the expensive stereo equipment that you will find in every college dormitory or student apartment. They speak of a world which is passing away. As Roger Palms points out in *The Jesus Kids* (Judson, 1971), their most quoted biblical texts have to do with the end time, the imminent disappearance of the world as we know it, the possibility of the "rapture" of the saints — all those apocalyptic passages that we rational, mundane Christians tend to dismiss as cryptic. They do not believe in the myth of progress. Their sexual mores are not "puritanical," but they do not consider sex highly important. They seem simply to have turned off the world of advertising and its glorification of sex, and they are grooving (as they would say) on essentials: salvation, sanctification, prayer, witness, fellowship, and the work of the Holy Spirit in their lives. Indeed, they remind one of the real Puritans, who lived so deeply that sex and all physical pleasures, while important, were not the focus of existence. Here again the "Jesus kids" flout the technological society. Where many of us do everything in our power to enhance our own attractiveness and keep our youth, where we pursue life as though it were about to slip from between our fingers, these "kids" are just "simple." Obviously, pleasure is *not* what life is all about; so they give it little thought and less time. They do not "buy into" modern society.

Intellectual historians cite many reasons for the demise of the invisible world of God, which men made visible by building cathedrals, organizing society, developing commerce and fighting the crusades. The Protestant work ethic, religious pluralism, the rise of modern science and its attendant philosophies (Cartesian rationalism and skepticism), the building of cities — all these have been blamed for the disappearance of the "sacred canopy" of the Middle Ages. Probably these did play a part, but observers as different as Thorstein Veblen, Peter Berger and Bishop E. R. Wickham place the major blame on the industrial process. For, they say, this process made the workingman perceive the world as a cause-and-effect machine, chained him to a god more efficient and powerful than any possible extraworldly power, and uprooted him from the rural society in which he had a role he understood. The process itself — the means of production — did more to fragment that invisible world than all the Humes or La Places with their philosophy or science.

In any case, the Jesus movement people are rejecting that process. It is too time-consuming and demanding; it requires allegiances they will not make because they consider them idolatrous. Indeed, their often shrill denunciation of the churches issues from their conviction that we church people have sold our simple faith in Jesus for a mess of pottage; that is, for a production and sales system with a reciprocating consumer mentality which cannot allow us to seek first the kingdom of God and his righteousness. The jobs they do take in order to live are jobs that demand of them a minimum of loyalty and time. In contrast to standard-brand Pentecostals, at whose services there are frequent references to people "moving up" to better, more responsible jobs with bigger rewards, these youngsters keep only a tenuous relationship to education and job so that nothing can rob them of the time and energy they need to be active in the Lord.

That's where the Jesus kids are now. Not for them a meditative process that, presumably, would enable them to live sanely in our mad world. Nor are they interested in hitching machines to people to measure the quantity of their religious experience. Their approach to modern life is semieremitic or communal or Anabaptist in that it understands the world to be at least provisionally in the hands of the prince of the power of darkness. They can feel right in living with that world only by being marginal to it. The radicals of the late '50s and '60s talked about being "marginal men." The Jesus youth *are* marginal men. Whether they can keep on that way will be the real test.

Technology And Society

A CHALLENGE TO PRIVATE ENTERPRISE

By IAN K. MacGREGOR, *Chairman of the Board and Chief Executive Officer, American Metal Climax, Inc., New York, N. Y., Chairman of the ICC Congress Committee on Use of the World's Resources*

Delivered before the XXIIIrd Congress of the ICC, Vienna, April 17-24, 1971

W E ARE TO consider today an important aspect of this year's Congress theme. "Technology and Society: A Challenge to Private Enterprise." Our concern is the use of the world's mineral and energy resources.

Sir Solly Zuckerman's background paper gives a sound analysis of some of the new problems that confront business as well as government. Indeed, these problems confront society as a whole on this planet, for in the past decade mankind has become aware, rather suddenly aware, of three basic facts of its existence.

First, our planet is finite. Fresh water, the air we breathe, arable land and all the mineral elements and sources of energy are finite.

Second, we have discovered that after the marvelous medical advances of the present century the entended life span of man implies explosive population expansion.

Third, although the world's rapidly expanding population depends utterly on industrial technology to produce the goods and agricultural products required to meet its growing needs and desires, the world has now discovered that it must urgently learn how to avoid, or how otherwise to deal with, what Sir Solly calls the "deleterious secondary effects" of technology itself.

It may take some time before society in all of its national, international and functional organizations can begin to deal effectively with the problem of the environment *in toto*, including natural mineral and energy resources as a part of the whole.

It is clear that international business, through the International Chamber of Commerce should now begin to consider its role in solving the total crisis of the environment.

Business—both national business and international business—*must* play a major role in dealing with this fundamental problem of mankind. As Sir Solly points out, the technological innovation we will need in order to solve this total environmental problem will flourish best in a competitive

VITAL SPEECHES, June 15, 1971, vol. 37, 525-529.

atmosphere. It may be equally true that our free enterprise system will only be ensured of its survival if it does play a major part in this effort.

We must look to the Free World's "industrial/agricultural/energy complex" to provide both high technology and massive amounts of goods. In the countries represented at an ICC Congress this production complex is largely organized by private capital. Unlike that so-called "military-industrial complex" of which so much has been heard in some circles, this "industrial/agricultural/energy complex" cannot be budgeted away by popular or political demagoguery. It is the basis of our existence.

There is surely a unique opportunity here for the ICC. This organization, bringing together leading businessmen of the developed and developing world, must champion the integrated use of the total environment. It must stress its conviction that we can best cope with the pollution threat, the crisis in natural resources and even the problem of rampant population growth within the framework of a profit-oriented society.

The problems of pollution have caught the imagination of the public in our various countries. There is even some danger that too much attention in the short run may be given to hasty, emotional and unbalanced approaches to some pollution problems. Witness the popular assault on the SST, the supersonic transport, in the American Congress during the past month. This is acclaimed as a victory by the new environmental purists. It is regarded as a serious defeat or even a turning point by those who believe in the necessity of continuing technological progress.

The problems of natural resources are not so obvious to the public. In this Committee of the ICC Congress it is our task to stress that the world is heading into as severe a crisis in the use of natural resources as any it faces in the realm of environmental pollution.

The case of copper illustrates the extent and nature of this impending crisis. Copper has been a key metal in all civilizations. With the discovery and use of copper over 6000 years ago modern civilization as we know it started its evolution. Ever since, the leading users of copper were the leading nations. For example, Egypt's outstanding place in the dim beginnings of our history was in no small measure due to its sophisticated knowledge of the extraction, fabrication and use of copper. In today's electronic age the use of copper also connotes the level of technological sophistication that any country has reached.

In the United States today we consume about 20 pounds of copper per person per year. Western Europe and Japan use about 15 pounds. In all of Asia and Africa, on the other hand, the use of copper today is only about 3/10ths of a pound per person. No matter what degree of success the UN Second Development Decade achieves during the seventies, we know that the average use of copper in the developing world is going to increase in the years ahead. Indeed, by the year 2000, if all the world consumed copper at today's average rate in the United States, Western Europe and Japan, even under the extreme assumption of zero population growth over the next 30 years, the world would need annually 30 million tons of copper, a sixfold increased over the total amount of primary copper produced today. On the other hand, as is so likely as to be almost a certainty, if there are twice as many people in the world in the year 2000, but if the developing peoples by then consume only half as much copper per person as the developed world uses today, which is still a formidable target, and if we of the developed countries did not increase our per capita consumption at all, the world would still need 35 million tons of copper per year by the end of the century, seven times the amount consumed today.

There is a further dimension to this picture of expanding use of copper. The United States Bureau of Census recently estimated that average family income in the United States will double over the next 30 years, measured in *constant* purchasing power. By the year 2000 the average United States family would have a real income of over $20,000 a year instead of nearly $10,000 at present. Considering the unique role of copper in modern living, a standard of living twice as high as we have today might well imply a considerable increase above today's 20 pounds per capita. At 30 pounds per capita, and with some 300 million people in the year 2000, the United States alone might need nearly as much primary copper as the whole world uses today.

It is affluence that largely contributes both to the crisis in pollution and the pressure on natural resources. Therefore, any such growth of real incomes in the United States and other developed countries, added to increased urbanization of the larger populations, will certainly compound the problems that lie ahead.

Where is all this copper going to come from? Will copper become a precious metal? Can we look forward to rising standards of living if copper cannot any longer be used as it is today but only as precious metals are used?

The same questions must be asked for many other metals that are essential to our advanced industrial and electronic civilization.

Whether we look back over the 25 years since World War II, or whether we project forward 30 years to the end of the century, the picture is the same: rapid expansion in the use of the world's mineral resources.

World-wide use of iron ore has increased at least 2-½ times in the past 25 years. Use of magnesium metal is up three times, chromium nearly three times, primary copper use has doubled, nickel has nearly tripled, lead and zinc are up 1½ times.

Let us look ahead for even the next 10 years. The Chairman of U. S. Steel Corp. expects the present global use of some 550 million metric tons of raw steel to expand by 50 per cent or more by 1980. The World Bank is reported as estimating that consumption of primary and secondary copper, which was 5½

170

million tons last year, may exceed 9 million tons in 1980. The use of nickel, tungsten and molybdenum are expected to double in this decade. By 1980 we should be well on the way, as a result of both population growth and economic development, to the rates of use of minerals in the year 2000 which sound so astronomic today.

Without making extreme assumptions as to the rate of improvement of the economic condition of the peoples in the developing countries, today's world demand for many key minerals is expected to triple by the year 2000.

In energy we find the same prospect of steeply increased use. Even with an assist from all the nuclear capacity that we can hope to build in the next 20 years, and with continued heavy dependence on coal, petroleum consumption is expected to increase threefold in the next 20 years. By the year 2000, with double today's population, the world's total use of energy is projected at five times today's. We can hope this goal will be reached because human well-being has always equated to the consumption of energy.

Statistics on the rate of consumption of metals in any one year do not provide a full measure of the resource problem facing the world. Not long ago (at a conference at the California Institute of Technology in 1967 discussing the topic, "The Next Ninety Years") Harrison Brown tried to estimate the *cumulative* drain on the world's resources by the year 2000 which might result from a doubled population compounded by much greater affluence among the poorer nations. Referring to a possible world population of over 7 billion persons by the year 2000, and using figures of the *cumulative* amounts of raw materials which would have been used during all the years until the year 2000, Professor Brown spelled out an astounding picture of possible total use of resources in the last third of this century, on one extreme assumption:

"If by some miracle all these persons were to be brought up to the standard of living now enjoyed by the people of the United States, we would need to extract from the earth over 50 billion tons of iron, one billion tons of copper, an equal amount of lead, over 600 million tons of zinc, and nearly 100 million tons of tin, in addition to huge quantities of other substances. These quantities are several hundred times the present annual world rates of production. Their extraction would virtually deplete the earth of all high-grade mineral resources and would necessitate our living off the leanest of earth substances: the waters of the sea and ordinary rock."

Coming back to earth, to an earth we can more easily recognize, it is an inescapable fact that, whether we look to the developed countries or to the developing countries, whether we look to industry or to the consumer, metals will be used in the decades immediately ahead on a scale many times what we know today, and we stand today at a peak consumption of metals in terms of all previous history. Indeed, more minerals have been used since 1900 than in all time to that date.

171

The clear warning on the horizon, and not a distant horizon, is that mankind will shortly have to face heavy increased social costs both to control pollution and to overcome diminishing natural resources. These added social costs are going to put severe pressure on mankind's standard of living.

The trade-off between environment and natural resources is also going to cause many agonizing decisions for society. An interesting recent example of this was a conversation (in a Profile in the "New Yorker" magazine of March 20) between David Brower, former Executive Director of the Sierra Club and now leader of a conservation organization called "Friends of the Earth," and Dr. Charles F. Park, Professor of Geology and of Mineral Engineering at Stanford University. They discussed the trade-off between wilderness beauty and copper mining. Dr. Park said "Minerals are where you find them. The quantities are finite. It's criminal to waste materials when the standard of living of your people depends upon them. A mine cannot move. It is fixed by nature so it has to take precedence over any other use . . . Proper use of minerals is essential. You have to go get them where they are. Our standard of living is based on this." Brower held that copper was not a transcendant value in one of the few remaining great wildernesses. He said, "We have to drop our standard of living, so that people a thousand years from now can have any standard of living at all." Dr. Park replied "I am not for penalizing people today for the sake of future generations."

Dr. Park fears, and I think David Brower hopes, that continued economic growth is doomed on a finite planet.

We have seen in our lifetime how rapidly the world is using up its *high grade* mineral resources, constantly reaching down to lower grades to meet its present needs. This raises the question whether modern civilization and all progress may not be in jeopardy because our globe and its rock crust and mineral supplies are finite.

There are those who are pessimistic and one of them is Dr. Park, whose lifetime study of this field leads him to the conclusion "that a constantly expanding economy, keeping step with a growing population, is impossible because mineral resources and cheap energy are not available on this earth in unlimited quantities." "The affluence of modern civilzation," he says, "is indeed in jeopardy."

However, Dr. Park does not conclude that all is irrevocably lost. He believes that if the expansion of the population can be brought under control and "an equitable distribution of the earth's mineral resource's among all nations on the basis of cooperation" can be achieved, we might hope to maintain and even to extend the affluence of our world society.

We must admit that this is a big "IF." *If* we are not to fear for our affluence or face the end of our kind of civilization, the world *merely* needs to work out a sensible position on population growth and learn to deal equitably with its finite and ever scarcer mineral and energy resources. This is indeed a challenge!

We businessmen know that the world will not be able to

meet that challenge unless economic forces are allowed to play their natural role without being overwhelmed by political, administrative and other frustrations that will only add to total social cost. Both in meeting the problems of pollution and the problems of resource scarcity, *prices* and *profits* must be relied on as principal mechanisms of change and adjustment. I recommend this as the message of the XXIIIrd ICC Congress.

The essential role of economic incentives can be illustrated in five fields of activity relating to the resource problem: conservation, substitution, processing cost, discovery and ocean mining.

First, conservation. We are all familiar with the way scarce resources are economized on in one kind of emergency situation—war. Most of us can recall the wartime measures taken 25 to 30 years ago in Great Britain, the United States and Western Europe, for example, to utilize scrap of all kinds, to increase scrap collection, to recycle, to substitute materials for scarcer kinds of materials, to forbid any use of certain materials for other than highest priority uses, to evaluate total present and potential supplies of basic materials on the widest global basis and to allocate their use nationally among nations allied in the common cause and within each nation. Most of us have some idea, and some a very keen recollection, of the even more draconian measures to conserve and rationalize the use of scarce materials that had to be taken by the Nazi war machine within its blockaded continental area.

In peacetime, we usually depend on the workings of our market economy to achieve the same ends, over a longer term, and without the imposition of the arbitrary authority of the state.

Now, in peacetime, we are going to have to learn new ways to deal with the wastes of technology and affluence as well as achieving greatly increased conservation in the use of scarce mineral materials and greater re-use of metals.

This process is already at work. One good example is the headlong rush of American automobile owners into pint-sized cars—many of them of European manufacture, because Europe has always had to be less profligate of materials and energy than we in the United States, and Europe's manufacturers have gained a considerable technical advantage in providing comfort, safety and movement on the small scale. Perhaps this is a beginning of the lowered standards that we must expect as a result of the resource crisis. On the other hand, perhaps it is *not* a lowered standard, but a technological advance, providing the same product—transportation—with the investment of fewer materials.

In the same way, perhaps we can have our doubled affluence in the United States by the year 2000 and still not use more than 20 pounds of copper per person—or use less. One can hope that our technology is that imaginative, that successful.

The recycling of materials is another link between environmental and resource problems. In the minerals field, perhaps to the surprise of many, we have always had substantial recycling. At one extreme is the care with which gold, platinum and silver and other precious metals are recycled; indeed, they lose little volume down through the ages. Many other metals have always been recycled and the size of the secondary materials industry in the industrialized nations today is evidence of this. A special form of the recovery of mineral waste is, of course, the recovery of virtually all the world's cadmium and a substantial and increasing amount of its sulphuric acid from exhaust flue gases of mineral smelters. This also lessens pollution of the atmosphere. The recovery of selenium, palladium, iridium as well as precious metals from the slimes resulting from the electrolytic refining of copper is another example of a form of recycling.

Second, substitution. Market forces must encourage endless substitution among mineral materials. Seldom does it happen that there is wholesale substitution of one material for another; usually the pattern is the gradual erosion of one use after another for a metal and the addition of other uses—usually new ones which can bear higher material costs. However, in the resource crisis we now face there are no doubt going to have to be some startling substitutions over the next 30 years.

Third, processing. We must continue to provide incentives to improve mining and beneficiaton processes thus making possible economic recovery of lower and lower grades of ore. Here is an optimistic view of the potential in this area (by Julian W. Feiss of the U. S. Geological Survey):

"In spite of the accelerating drain on the world's mineral resources, it is now recognized that they will never really be exhausted. Just as advances in technology have made it possible to exploit today ores so lean they would have been considered worthless only 50 years ago, new advances will make it possible to extract metals from still leaner ores in the future. In effect, technology keeps creating new resources."

What has happened in the United States in the mining of copper over the past few decades will have to be repeated for many other minerals. If our economic system is allowed to work, it is probable that ores having one-half or one-third the copper content of the grades presently required for commercial operation will be mined profitably before the end of the century.

Fourth, discovery. We will have to devise new incentives capable of producing an almost unprecedented rate of discovery of new mineral reserves. Evermore ingenious modern exploration techniques will find evermore elusive minerals—*if* there are profits to be made at the end of the search.

Finally, the ocean. Beyond the traditional earthbound search for mineral deposits that are harder and harder to find and of lower and lower grade, there are the oceans. The impact of increasing values for essential minerals may speed up the research and eventually solve the current economic problems of ocean mineral search and recovery, but this will only be

174

after high expenditures on research and development of means for economic mining and extraction in the deep sea environment. I must warn against over-confident anticipation that the solution to the minerals problem will be found in any large scale new access to ocean metals within the time dimension we are considering, that is, by the end of this century. There are severe economic technical and institutional obstacles to be overcome before marine mining passes beyond the research stage even into early infancy as an industry.

Unless the opportunity for profit at the end of a long period of research and development expenditure is clearer than it appears to me to be today, private capital is not likely to be harnessed to the effort to overcome the massive technical obstacles that stand between present technology and economic methods of mining the deep sea.

Major nations, including my own, and the United Nations, now appear to be agreeing to virtually internationalize the deep sea beyond the 200 meter depth. However, a very small part of the ocean's mineral resources are contained in the area of the continental shelf up to 200 meter depth. It is doubtful whether venture capital is likely to be attracted to the development of extractive methods for such internationalized ocean depths, in the context of current concepts for the control of those areas.

Petroleum, of course, is in a different category. It has been taken from the continental shelves for more than 30 years and today I believe as much as one-sixth of world petroleum production is obtained from off-shore sources. Given oil's unique fluid properties, deep sea extraction technology is apt to make faster progress than in the case of hard minerals.

Economic incentive is the common thread in all five aspects of resource use briefly reviewed above. As a noted economist (Professor Kindelberger of MIT) has said, "it is impossible to run out of anything in economic terms." With greater scarcity or greater demand, prices rise and this either encourages greater efforts to produce the goods in question, or it brings about the use of substitutes. We must hope that such simple— and profound—truths will be accepted as the basis of policy making as society begins to deal with its resource crisis.

The significance of this for our consideration today is, I believe, the following: Where there are market means for establishing a value on a product that is scarce or would otherwise be wasted, then recycling or conservation or substitution is automatically achieved by our profit-oriented economic system. What society must now do is to find new ways of giving value to such scarce or waste materials and new incentives for process improvement and exploration. We should study how new systems of pricing can achieve this, using cost and profit as motive forces, even if the new systems are established by governments.

The recent Annual Report of President Nixon's Council of Economic Advisers presented a valuable analysis of how "prices" can be set for the use of air and water based on the units of pollutants discharged, including variable subsidies for control of pollution, charges for polluted emissions and sales of transferable environmental usage rights. We must have similar innovative thinking about the use of new pricing techniques to solve natural resource problems.

We businessmen who believe in the efficacy of the free market system dislike authoritarian intervention in the market system. And we know that, in the end, the market system can cope with most shortages, can bring about massive substitution of one material for another even in a very short term, and can bring about a shift in the flow of resources as between users or industries and even between nations.

What we have to consider now is whether it is wise for us to allow the market to do its work without our trying to gauge where it is likely to take us and then, in the light of such an assessment, possibly accepting certain overall governmental measures of preferably indirect scope in order to influence or to speed up those trends which we find either inevitable or desireable.

In one way or another many governments and the world community through the United Nations, the OECD and other organizations are engaged in some aspects of such work. It would be appropriate for international business through the ICC also to begin to consider ways of dealing with the problem of diminishing grade and quantity of the ores from which most of our industrial raw materials are derived.

In the crowded center of Western Europe it has become all too painfully obvious that pollution of air and water has no regard for national boundaries. Mineral and energy resource problems have their own way of paying little respect to national boundaries. For one thing, the industrialized countries are now heavily dependent on supplies from the less developed countries, just as the latter begin to increase their own use of scarce materials. Their more rapid rate of development in the decades to come will complicate the problems of our own rising use of materials. The less developed countries are also heavily dependent on their foreign exchange earnings from the natural resources they have the good fortune to be able to sell to the industrialized nations. Therefore, we must now begin to consider how the world's resources are going to be apportioned 30 years from now.

Unless the developed countries are willing to pool their knowledge and skills in assessing their future needs for mineral resources and working out a rational means of assuring their needs, they will find themselves captive to powers with strange priorities. This assessment will be essential if the developed countries are to continue their own progress. It will also be essential if they are to be able to offer any help to the developing countries in raising standards of living and achieving stronger political stability in the two-thirds of the world which face the principal problems of poverty and population. If the developed countries do not consider how they can expect to obtain the mineral and energy materials with which to support sound economies in the

176

future, they have little hope for continued stability and there will be no hope at all for stability among the less developed countries. They may have some most important supplies of minerals, but the developed countries have access to some too, and, more importantly, they have the capital, the skills to improvise and substitute, and above all, they have the huge markets.

The moral of all this for our XXIIIrd Congress of the ICC is that both developed and developing countries must work toward better ways of dealing with their own future mineral resource and energy needs. Something should begin to be done about this now. The ICC should be a central point for this new consideration of the essential needs of our society—and of international business.

One must have faith that future international consideration of natural resources will be practical and that it will be sensibly oriented towards economic, financial, technical and geological realities. One can hope that the forthcoming international Conference on the Human Environment in Stockholm in 1972 will give significant attention to the supply of mineral and energy resources necessary to support sound economies in the various nations of the world as they begin to grapple with the improvement of the environment. The ICC should certainly organize its work to support such considerations in Stockholm. I also recommend that the ICC work through the national committees of BIAC (Business and Advisory Council to OECD) to press for consideration of natural resource problems within the OECD.

How to Love the Land and Live With Your Love

Ruth C. Adams

Going South on Route 100, you turn East at Bucktown on Route 23. Driving through five miles of rolling countryside, fine old stone houses and woodlands, you notice bake ovens, smoke houses and tall whirling windmills. Windmills? You cross the boundary into Lancaster County, Pennsylvania, and you are in the eighteenth century.

A vast panorama of magnificent farmland laced with winding lanes opens on both sides of the road. Sturdy, timeless houses and great stone barns nestle in cozy groups, with chicken houses, stables, sheds, poles sprouting huge, old dinner bells. And windmills. The fields are almost unbelievably verdant and fertile. Sculptured in greens, browns and reds, they lie on the lavish landscape wantonly, promising rich, rich harvests in the season ahead.

There is something peculiar about these barnyards. There are no cars, no trucks, no television antennae, no power-driven machinery of any kind. There are no electric power lines marching down these lanes. Are these people too poor for such amenities? No. These are the old Amish. These are the Plain People who live, as they wish to live, departing hardly at all from the way their ancestors lived in the times of the Reformation. The House Amish, the Plain People use no electricity. They own no cars or trucks.

By the time you get to Morgantown you know how they get along without them. You meet shiny, black buggies and wagons pulled by fine, well-tended horses. The Plain People are riding in open or closed buggies or small wagons. If it's Sunday they are visiting friends, relatives or neighbors, or possibly coming from church, which is held in private homes rather than a church building. On weekdays the roads are almost empty, for the Amish are at work in their fields, barns and households. Most of them raise all of their own food. They butcher their own hogs and cattle, milk their own cows, pick their own apples, harvest their

Reprinted from Jerome Goldstein (ed.), THE NEW FOOD CHAIN: An Organic Link Between Farm and City, 1973, Emmaus, Pa.: Rodale Press, Inc., pp. 36-41.

own grain and take it to the mill to be ground. Their luxuriant vegetable gardens boast rows and rows of corn, peas, lettuce, beets, carrots, beans, kohlrabi, brussels sprouts, spinach. An Amish housewife may have 22 vegetables in her garden.

Inside the house, furniture is staunch, well-made, unadorned. The kitchen is the center of family life, the only room heated in winter. An old-fashioned black wood and coal stove provides for heating and cooking. The spinning windmill pumps water into a cistern from which the Amish housewife hand-pumps it into her sink. All the Amish clothing is made by the mother and daughters of the family. It is designed as Amish clothing has been designed since the Reformation. Little girls and their mothers wear long dresses with full skirts and aprons. The bright colored cloth is always plain, never figured. Aprons for everyday use are black. For Sunday and special occasions they are white. The men wear black. Everyone has a Sunday-best outfit. The rest is strictly for everyday. Women wear prayer caps at all times, their long hair in tight, neat knobs at the back. The men have hair to their shoulders and, as soon as they marry, grow beards. Their broad-brimmed black hats (straw for summer) are handsome and lend a dashing air. Boys of four wear the same kind of hat with great dignity. Little girls and women may wear bonnets. Clothes are practical, comfortable, made of easy-to-launder cotton.

The refrigerator in the kitchen may be run by bottled gas in some of the more progressive homes. An old-time refrigerator might be a handsome chest-like affair in the corner, with pipes bringing fresh water from the stream, which pours down through the double walls of the refrigerator, cooling it to 50 degrees or so even throughout the summer. The water runs off into the barnyard for the cattle to drink. The pump at the stream may be an elaborate but simple-appearing gadget which looks as if it runs by perpetual motion. There is no electricity or other power source. Elaborately-balanced water wheels keep the pump working away day and night, so long as the stream runs.

Nothing is wasted in the Amish household. Scraps from sewing bees are worked into quilts for warmth and decoration. Table leavings and waste from the household canning are fed to pigs and chickens. An Amish housewife may can or preserve as many as 1000 jars of fruit, vegetables, pickles, meat. She smokes bacon and ham in the smokehouse. She makes sauerkraut in a big tub. She makes apple butter in a huge kettle.

Clothing is worn until it wears out, then replaced with another piece exactly like the one which wore out. Fashion is never involved. Baths are simple affairs in metal tubs or crockery basins. Nowadays a flush toilet and a bathtub are permitted in some districts, if they are located downstairs so that no electric power is needed to carry the water upstairs.

The Amish are conscientious objectors and participate in no wars. Their hatred of the military is so great that they shun buttons on their clothing because ancient military uniforms displayed rows upon rows of buttons.

Today the Amish in some parts of the country are in trouble with the law. They are convinced their children should not go to school beyond the eighth grade, for they say children do not need any education except for the education that comes from living with The Land. They will be farmers always. They have no other occupation. Higher education is not only unneeded, they say, it tends to corrupt the farmer, perhaps turn him from The Land, which is all-important.

The Amish have an almost obsessive ferocity about The Land. Every waking hour, every thought is given over to the soil and the good things that come from it. They quote Bible references to explain this devotion. Some people find their outlook narrow, their conservative, highly moral religion stifling. And some young folks do leave the Amish community and become "worldly". The number is very small—perhaps two percent.

The Amish have always lived with The Land, in perfect harmony, returning to it everything they take from it, cherishing it as the source of all life. Living with The Land is hard work. You get up before dawn; you work till dark, then you light a candle or an oil lamp to finish the chores. You have the immense satisfaction of seeing the immediate results of your labor in the harvest, the cold cellar full, the can cupboard full, the corn-crib full, the hay-mow full—all by the first frost.

Late fall and winter are for socializing in those hours when you cannot spread manure, mend harness or prune trees. Social activities are determined by the distance your horse can go in an hour's travel or a day's. Although most Amish read newspapers and magazines, they like best their own newspaper, *The Budget,* printed in Sugarcreek, Ohio which is a chronicle of the doings of the 50,000 Amish all over the country. The December 16, 1971 issue carried news like this: "Dayton, Virginia: It doesn't seem like winter as the temperature was in the 60's." "Meyersdale, Pa.: Grass is nice and green and we had our cattle out in pasture most of last week." "Spartansburg, Pa.: Mennos had supper at Dave D. Byler's Thursday evening. They took a load of their belongings in to Geauga Friday and will take another load Tuesday and the rest Wed. They will live in Dan J. P. Miller's small house until they find something else. We are planning to have a singing for them tonight."

News like this is all the 14 pages of *The Budget* contain, except for a few ads for the very few products an Amish man might buy once in a lifetime—a clock, a blanket, a few yards of cloth. One who customarily listens to the six o'clock news of mayhem, disaster, murder, rape, addiction, war, graft, robbery, laced with idiotic commercials for trivia, can be excused for wondering if it might not make for a better life to have no daily news but *The Budget.* These simple family notes are somehow almost all that are required by people who have renounced all wars, all violence, all commercialism, and, it now becomes evident, in an environmentally concerned age—all pollution.

Except for a bit of cloth, enough fuel to heat one room, some few tools and enough hardware to rebuild their wagons and plows, the Amish need almost nothing from outside their communities. So far as their way of life is concerned, we could close down most of our major industries with all their pollution and scandalous waste of resources, and go back to The Land to live. Nor is there any reason to be poor. The Amish have done such an excellent job of living with The Land that, for a considerable number of years out of the last 100 years, they have been listed in Department of Agriculture statistics as having the highest farm income in the country.

They obey all federal regulations about restricting crops, but they refuse to accept subsidies. They refuse social security, too, and have no need for it, as their old folks are well cared for and highly respected as the wisest people in the community. And they are, of course, since they know more about farming than younger people, and knowledge of farming is the only knowledge demanded of an Amishman. No one feels any need for insurance as neighbors will help rebuild a barn which burns, relatives and friends will provide for a widow or orphan, without question or remuneration.

The struggle of an Amishman named Yoder against the might of the State of Wisconsin which has decided he must send his children to high school or go to jail has now reached the Supreme Court. In the January 15, 1972 *Saturday Review* you can read the full story. Says Stephen Arons, "Perhaps the greatest irony of the case is the notion of freeing Amish children from their community. The Amish community experiences little delinquency, causes and fights no wars, uses no polluting machines, eschews materialism and has no economically-based class system. To save these people from the quiet sanity of their lives by forcing them into the center of the psychologically unhealthy atmosphere of modern America strains the definition of freedom beyond recognition."

And the prosecutor, John William Calhoun, speaking for the State of Wisconsin, says, in part, "The issue is not the Amish life-style or a matter of pluralistic vs. egalitarian societies. In fact, to many of us caught in the remorseless, day-to-day crunch of daily living, the Amish life has great appeal. However, as Stuart Chase has said, 'Retreat to a simpler era may have had some merit 200 years ago when Rousseau was extolling the virtues of the Cro-Magnon man, but too much water has gone through the turbines.' "

Has it? Are we going to take the word of one economist and one lawyer that it has? You can live with The Land as the Amish do, if you care enough to do it. You can buy a windmill and an oil lamp in Lancaster County. You can get a horse and buggy, a plow and a cultivator. It is, apparently, perfectly possible to live a successful, fulfilling life with The Land, without any of the blessings of our great commercial-industrial establishment.

The Amish have been demonstrating to us for hundreds of years that all you have to do is care enough about The Land, to cherish it, to regard it as the single most important thing in life, the thing for which you are willing to give all the hard work and devotion necessary to make your relationship with it successful and permanent.

FUTURE PERSPECTIVE

Technology in the United States: The Options Before Us

J. Herbert Hollomon
Director, Center for the
Study of Policy Alternatives,
School of Engineering,
M.I.T.

In the first installment of this essay (*see "Issues for the 1970s* in Technology Review *for June, pp. 10-21*), we have identified and described a number of problems relating industrial progress, research and development policy, and scientific and engineering manpower which now confront the U.S. Briefly summarized, these observations are:

☐ The economy of the United States has evolved from agricultural to industrial to service-based. Past improvements in productivity have come largely from the agricultural and manufacturing sectors.

☐ The growing and widespread social consequences of industrial activity and the use of certain products have only recently begun to receive significant technical attention or government action and must be considered in the future industrial development of the society.

☐ As technology has spread throughout the world, competition from overseas has grown and can be expected to continue. The growth of the Common Market in Europe and the World Market for Japan gives to each of these economic units many of the advantages that the United States has enjoyed uniquely in the past.

☐ The system for educating scientists and engineers in the U.S. has been geared to meeting an ever-growing demand, largely based on the growth of space and defense programs. Recent decreases in their support has led to unemployment and declining salaries and will continue to do so unless other actions are taken.

☐ The prices paid for scientists and engineers have been inflated significantly more than other salaries and wages in the economy. The cost of all scientific and technical activity, whether aimed at increasing industrial productivity, improving technical capabilities, or dealing with social problems, has increased out of proportion to other costs.

☐ While support for research and development to improve health ser-

TECHNOLOGY REVIEW, July/August 1972, 32-42.

vices and aviation has grown, total public expenditures supporting research and development for education, the criminal justice system, non-aviation transportation, health care delivery, and the disposal and treatment of waste are almost insignificant.

☐ Increases in productivity do not come directly from research and development alone; they involve experience in manufacturing, the supply of services, the diffusion of old technology, and public support for a social climate that encourages and adapts to change.

☐ There is a good correlation between industrial growth, productivity, and investments in research and development for many industrial activities; the correlation is less good for the electronics and aviation industries, which may be less effective in exploiting research and development than other industries that received less governmental support.

☐ Second-order indirect social costs of technological change have seldom been considered in the calculations of its costs and benefits.

☐ Recent studies indicate that research and development expenditures correlate positively with profitability, but the correlation is much less certain than indicated in studies made in the early 1960s; the profitability of research and development may have declined.

☐ Large investments in research and development are typical of growing industries and may contribute to their growth and profitability. Less dynamic and older industries support relatively less research and development, and this may further depress their growth.

☐ The primary processes of technical change, at least in relation to civil activities, may depend less on new research and development than on ingenious applications of old techniques in response to market demands.

We concluded that these issues in relation to present U.S. social problems make clear the need for revision of U.S. policies relating to technology and its use in the society. But we cautioned that our analysis of past policies also makes clear the need for a better understanding of the effects of research and development policy as we analyze future alternatives.

The several policy alternatives that follow are an attempt to enumerate possible courses from which we might choose in order to make more effective use of technology in society. Even though the studies and analyses of past policies are inadequate, our present situation clearly demands the consideration of immediate action. The options presented are discrete; their possible interrelationships in combinations of two or more have not been considered here, although they would have to be considered were they to be proposed as federal policy. Some options exclude others, some do not.

Option 1: Take No Specific New Actions

One policy option is to allow present trends to continue and to take no new policy actions aimed at making technology more effective in our society. A continuation of past policies will lead to continued and growing federal support for research and development as well as support for technical activities to improve the delivery of public services (such as health, education, and transport) and alleviate societal problems (such as crime). This kind of non-defense, non-space activity has grown during the past ten years at the rate of approximately 12 per cent per year, and we will assume a continuation of this growth for the future.

If we pursue the policies of the recent past, the relative decrease in research and development related to military activities will not occur as rapidly in the future as it has recently, and the percentage of G.N.P. devoted to research and development will level off at approximately 2.7. We assume then that the growth of industrial research and development as a fraction of G.N.P. will continue as it has for the last decade or two; no new conflict will generate new technical demands for hardware, and real per capita G.N.P. will grow approximately 2.5 per cent per year.

This set of assumptions obviously is based on a continuation of present policies that deal with the social consequences of technological change, a renewal of economic growth, and a continued decline in military commitments. Under these conditions production increases will result largely from the re-employment of the physical and human plant that is now partially unemployed. The productivity of the services sector will not increase substantially. The overall decline in the demand for technically trained manpower in comparison with its supply will continue for the next few years. As a result of excess supply, the prices paid for scientists and engineers will continue to fall relative to other wages and salaries. The number of college students who opt for the physical sciences and engineer-ing will decline, eventually causing a decline in the supply.

Based upon the assumptions made about the national economy and using a supply/demand model for scientific manpower, Richard B. Freeman has predicted the number of trained people that will be produced in the future. His model predicts a substantial change in the supply of scientists and engineers in the late 1970s. Allan M. Cartter (who both vastly overestimates the supply and underestimates the demand) and others have also presented analyses of future trends in the supply and demand for scientists, but their analyses do not consider the delayed, yet significant, response of students to changing market conditions. The chart on page 35 illustrates the difference between the estimates made by extrapolated trends and those derived from the model incorporating market response mechanisms. Consideration of market phenomena leads to a prediction that one-quarter as many Ph.D. physicists will be produced in 1980 as are predicted by trend-extrapolation techniques. About three-quarters as many doctorates in engineering will be granted in 1980 as were granted in 1970. An equilibration of the supply and the then-growing industrially- and publicly-supported demand is predicted by approximately 1975. At that time, the relative prices for scientists and engineers and the cost of research and development will stop declining. Industry will have increased its commitment to all kinds of technical activities as a result of reduced costs. Although the supply of science and engineering graduates will be roughly equal to demand, a far smaller number will be graduating in 1975 than in 1970. Following 1980 the decrease in the college-age population in the United States may lessen the demand for scientifically and technically trained manpower in colleges and universities, but this will be partly offset by the growing demands in other sectors.

The positive consequence of this policy alternative would be an increase of 25 per cent to 30 per cent in the overall technical activity in the United States by the end of the 1970s. This increase, largely the result of the reduced cost of scientific and technical people, should lead in the long run to some improvement in productivity, in the supply of new products, and in the effective use of technical people to moderate the social consequences of technological change.

The negative aspect of this alternative would be its failure to decrease the current unemployment of scientists and engineers, resulting in an adjustment period of four to five years in which the potential contributions of unemployed, or underemployed, technical people would not be realized. The pool of scientists and engineers would decline not only by attrition but because of the high obsolescence rate of their skills if not continually used.

Furthermore, this alternative would allow no *new* major commitments to pressing social problems, such as the improved delivery of services or the amelioration of the indirect effects of technology employed in the past. Neither would this alternative correct the underinvestment inherent in innovative activity by industry, nor would it create any additional activity aimed specifically at alleviating the growing disparity between foreign and U.S. technical capabilities in non-defense, non-space pursuits.

Option 2: Directly Support Private Technical Efforts

Economists have long known that in competitive free markets the single firm cannot capture all the benefits of its innovation, though it must bear the major costs. Furthermore, if the cost of the innovation is high compared with the financial capability of the firm, the risks, related either to possible failure in the market or to uncertainties about the success of developing the technology, may be too great for the firm to accept. There are also social and political obstacles to innovation related to the acceptance of new products or processes and the social and human adjustments that must occur as a result of the innovation.

In the last two decades the rising costs (salaries) of U.S. scientists and engineers documented previously have so raised the costs of innovative activity in private industry as to considerably deter it. The inflated salaries and the decrease in the growth of technical activity may have been a factor in the decreasing rate of productivity increases in the private sector; productivity increases in the period 1966-71 were about half that per unit input of the preceding two decades. Inflated salaries certainly contributed to the current situation in which Japan can employ two to three times as many scientists and engineers as the United States for an equivalent expenditure and, therefore, can effect technological innovations at lower relative costs than the United States while taking advantage of U.S. technology through the purchase of patent

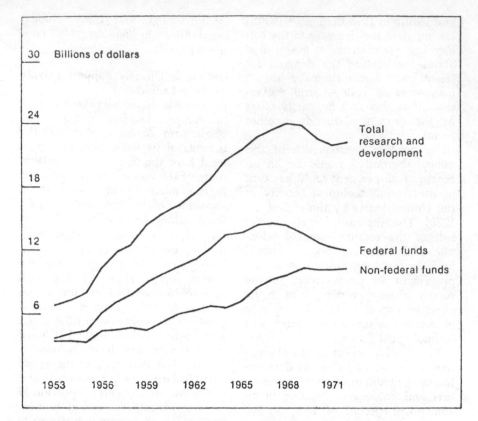

30   Billions of dollars

24

18

12

6

Total
research and
development

Federal funds

Non-federal funds

1953   1956   1959   1962   1965   1968   1971

Two changes have affected the pattern of U.S. research and development in the past two decades: the steady rise in total research and development investment (in constant dollars) which many scientists and engineers began to take for granted in the 1960s has in fact ended; and the proportion of our total national effort due to federal funds has declined. Do these trends, continuing into the future, augur an inadequate technological base for American industry? (Data in 1967 dollars: National Science Foundation)

rights and know-how.

All these factors suggest a national policy and program that would reduce the costs to the private sector of invention, innovation, and diffusion of technology. Whether the private sector produces the goods and services that best benefit society can be viewed as a separate question. Separate policies can create incentives and disincentives to alter the direction of industrial activity. The latter can be achieved through such means as pollution controls or the creation of a market for new public services; but whatever the direction of industrial activity and the social, political, or economic goals, the processes of invention, innovation, and the diffusion of technology must accompany them.

One mechanism that would reduce the costs of innovation to the private sector is a direct subsidy or tax credit for industrial research and development. A 35 per cent subsidy of industry's research and development costs would simply return the costs of technical activities relative to other costs to the level at which they existed prior to the major distortion introduced by the large federal research and development involvement of the 1950s and 1960s. Since all technical salaries have been inflated and since research and development is only one of the modes of innovative change, this size of subsidy of a firm's total technical effort could theoretically be justified. Obviously, the subsidy would stimulate the demand for scientists and engineers and maintain the high prices now paid to them. Indeed, the fact that the benefits of innovation are not fully appropriable might argue for the maintenance of a small subsidy for all innovative technical work; however, this is probably politically impractical.

Such a subsidy could be provided, either for all research and development investments made by a firm or for incremental research and development investments above a certain historically determined base. The latter proposal more reasonably meets the argument that research and development now supported by firms is economically justified and should not be subsidized. It has the advantage of leaving investment decisions closely coupled with market conditions and tends to correct the inflation of costs now present in the entire industrial system with respect to technology. However, the size of the subsidy must be carefully considered. Too large a subsidy would encourage high salaries as it increases employment.

Audits and Controls

The arguments against such tax credits center on anticipated administrative difficulties rather than on questions of economic principle. Any subsidy would encourage individual firms to call many industrial activities research and development in order to reduce costs and increase profitability. This difficulty might be overcome by recognizing that increased research and development must be accomplished by employing more scientists and engineers, an action which certainly could be subject to audit. In addition, data collected by the Bureau of Census and the National Science Foundation indicating long-term trends could serve as a basis for judging whether increases in research and development were in fact stimulated by the tax subsidy. Since about 90 per cent of all research and development in industry is conducted in 300 large firms, the policy of tax credits would not seem impossible to implement, although one function of the tax credit would be to encourage smaller firms to engage in research and development.

Some argue, too, that the tax system should be concerned only with collecting revenues and not with correcting difficulties inherent in the economy. Others argue ideologically that federal subsidies should be based on political judgments of what is "good" for the society, that the issue should not be left to forces of the market and public regulation. It is clear that a tax subsidy will not necessarily encourage investments in those activities that deal with the broad social needs of the society. These needs will have to be supported directly or stimulated separately in private industry by impos-

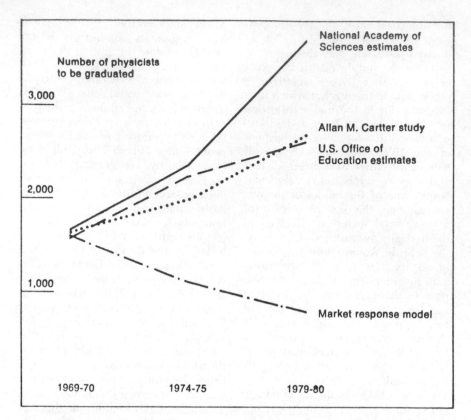

Number of physicists
to be graduated

3,000

2,000

1,000

National Academy of
Sciences estimates

Allan M. Cartter study

U.S. Office of
Education estimates

Market response model

1969-70   1974-75   1979-80

Though estimates of the supply of new Ph.D. graduates in physics based on extrapolating current trends differ widely, they all differ even more markedly from estimates made with a model which incorporates market response mechanisms.

In other words, we often fail to realize how sensitive are the career plans of college students to their perceptions of future needs and rewards in any particular profession or field.

ing specific incentives or disincentives. This argument implies that the policies for encouraging the *processes* of technological change can and should be separated from the policies affecting the *purposes* to which the processes are applied.

Innovative industrial activity can be stimulated by other means as well. The technical base—i.e., the state of the art—on which technological innovation takes place in industry could be, as it is now in part, supported directly by government. Two activities basic to improving industrial output are the development of the production process, which involves automation, management, and the design of production equipment; and the development of design methods and techniques for new products. These activities could be supported directly in technical schools and universities as investments in the technical base of the society and as a way to influence the training of young scientists and engineers toward concern for industrial problems. This support would be similar to that now provided by the defense and space agencies which, to encourage the advance of certain technologies, subsidize research and the training of people within specialties basic to their missions.

Support for Basic Science and Engineering

There has been little question that certain scientific and technical efforts were necessary to develop the resources subsequently used in the space and defense effort. Analogously, it might be desirable to provide support for the non-appropriable work required to sustain the technology and science underlying industrial innovation. This support, primarily through grants to universities, would reduce the technical risks of innovation and would provide new, effective couplings between universities and industry. Perhaps grants to universities might be restricted to those cases that offer some assurance of industrial cooperation, possibly through associations or through matching grants by industry. The potential effectiveness of this mechanism can be supported by studies of defense-related innovations, which clearly indicate that a disproportionate number of the individuals involved in defense innovation came from those schools that received large amounts of defense research support.

One way to estimate the relative size of such a support program would be by considering the ratio of support given universities by de-

fense and space agencies to the agencies' total research and development activities. This ratio was 3.4 per cent for defense and 3.3 per cent for space in 1969. Since the total of industrially-supported research and development is roughly $12 billion, the level of support given universities might be on the order of $400 million. To best connect university activities with industrial needs, it might be desirable to establish a program in which both government and industry participate.

Basic civilian science and technology might also be encouraged by establishing a series of government- and industry-supported research institutes coupled to universities. These would be similar to the Max Planck Institutes in West Germany, which were largely responsible for the great scientific strength of Germany in the early part of this century and appear to be a significant means for closely coupling university science and industrial technology in West Germany today. While Great Britain's public support for industrial research associations has been criticized, the United States might explore the underlying idea of connecting the universities to associative activities. Most industrialized countries other than the United States have used this mechanism to achieve the diffusion of technology; the United States might find this kind of association an effective mechanism for improving the capability of the vast majority of small firms that cannot afford to perform their own research and development.

Option 3: Indirectly Support Private Technical Efforts

As indicated previously, industrial

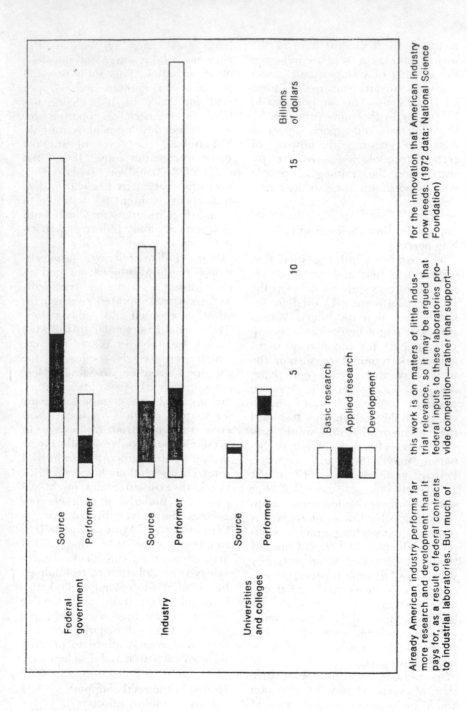

Already American industry performs far more research and development than it pays for, as a result of federal contracts to industrial laboratories. But much of this work is on matters of little industrial relevance, so it may be argued that federal inputs to these laboratories provide competition—rather than support—for the innovation that American industry now needs. (1972 data: National Science Foundation)

innovation appears to be most successfully encouraged by the "pull" of market demand. New agglomerations of markets and increased demand encourage investments in research and development by reducing market risk, while declines in demand often retard investment in technology. There is little doubt that the low investment in technology in such U.S. industries as shoes and textiles is related to their relatively slow growth. Other factors that characterize a conservative, change-resistant enterprise are probably involved in the construction industry's failure to capitalize on research and development: restrictive labor practices, product codes, and standards aimed at protecting vested interests.

It is also true that—particularly in housing—firms are often small and unable to undertake the high-risk technical activities required to bring about rapid product improvement or significant efficiencies. Ezra Ehrenkrantz has demonstrated that when the individual requirements for a number of new school buildings in California were consolidated into a single performance specification, industry responded with innovative ideas that permitted the construction of more efficient buildings.

Textile and shoe manufacturers could apply this lesson by agreeing to set performance standards for radical new machinery; without such radical technological change, these industries may be unable to compete with foreign firms paying significantly lower wages. With encouragement by federal subsidy, research and development might be stimulated by the prospect of a new, large market for shoe and textile equipment. Increased domestic production of advanced machinery would have the additional effect of reducing the present importation of machinery and might even stimulate an export market. As a large purchaser of civilian goods, the federal government could agree on performance specifications that require industrial innovation and pay for the prototype development to reduce both market and technical risks to suppliers. A program of this type was initiated several years ago by the General Services Administration for the development of government administrative buildings and could be extended to food products, clothing, medicinals, and any goods the government buys in large quantities.

Techniques developed as a result of this mechanism should be directly applicable to the production of goods for the civilian market. Evidence of the effectiveness of this technique of "pulling" technology has been established by computers, airplanes, and integrated circuits produced originally for the federal government that now have extensive commercial markets. The federal government could extend this mechanism by requiring, for example, that hospitals receiving federal support be constructed through cooperative efforts that set performance standards for successive buildings.

Other indirect means exist for encouraging technical development in firms or industries with little knowledge of modern techniques. A novel notion now being tried in Canada is to support the education of graduate students partly through direct grants and require them to work in industry for the remaining support for their graduate activity. Such a program might stimulate the industries themselves to support other people.

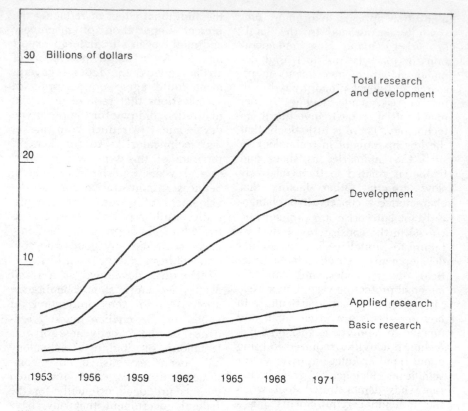

30 Billions of dollars

Total research
and development

20

Development

10

Applied research

Basic research

1953   1956   1959   1962   1965   1968   1971

What should be a rational balance between basic research, applied research,
and development, in a total national research and development program?
Though there is no certain answer, the
balance has not significantly changed in
the past two decades in the U.S., despite
considerable change in priorities and in
the technologies to which research and
development is devoted. (Data: National
Science Foundation)

Students, in turn, would be stimulated to be knowledgeable about the
problems of the industry supporting
their education. Government could
receive benefits similar to industry
by initiating apprentice programs
within government agencies.

Option 4: Improve the
Services Sector
With half of our workers engaged in
providing services, the United States
has become the first post-industrial
society. The fastest growing services
are health, education, and local and
state governments. These services
are provided by a large number of
diverse establishments, few of which
are able to support the technical efforts necessary to improve their efficiency or effectiveness. The fire-
fighting, police, welfare, road building, and sanitation activities of local
governments, for example, have not
benefited significantly from advances in technology.

The services sector might be likened to highly fragmented industries; but lacking the discipline of a profit motive. Technological innovation is discouraged in the institutions that supply education and health services, for example, by their structures and incentive systems which inherently discourage cost-reducing changes. In these institutions, the individuals who determine the system's operating characteristics and level of effectiveness—for example, doctors and teachers—do not have much incentive to reduce costs. Indeed, they often find it convenient to increase delivery costs, since they are not forced to bear any of them directly.

Yet there are technologies basic to each of the several components of the services sector which could, if used, improve the effectiveness of the services they supply. Community health plans that associate medical facilities with treatment and provide financial incentives to reduce health care costs may be one way of restructuring the health delivery system to encourage innovation and improved service. Although no equivalent scheme for improving either schooling or government services has yet been tested, the soaring costs of services and decreased growth in demand for them will begin to induce a climate for innovation to improve their effectiveness and efficiency.

Improvements in government, health, and education services involve invention, innovation, and diffusion as do the other sectors of the economy. These processes must be supported either directly by the government or by individual institutions under incentives to support the innovative process themselves. Cur-

rently, the percentage of research and development allocated to improve public services is small compared with the total costs of providing those services. Correlations between productivity improvements and research and development in industry may serve as a basis for estimating the amount of research and development that could be justified for health and education services. Growing and profitable industries devote at least 4 per cent of their sales revenue to research and development. A level that is 3 per cent of the total health and education services expenditure would require a public research and development investment of $5 billion. It must be remembered, however, that no amount of research and development, invention, or even preliminary innovation is sufficient to ensure that changes will be adopted broadly and diffused throughout the services sector; the institutions themselves must accept the changes that are devised.

As in manufacturing, there are techniques connected with the storage, manipulation, recovery, and analysis of information that are basic to each of the components of the services sector: in education, the science and technology of learning and teaching; in health care delivery, the technologies associated with testing, diagnosis, and prevention of sickness (rather than the cure of disease); in government, the application of operational analysis and control procedures to such activities as fire-fighting and the allocation of police resources have already been demonstrated. These techniques could be developed for the services sector by the universities with the support of research and development and the encouragement of coopera-

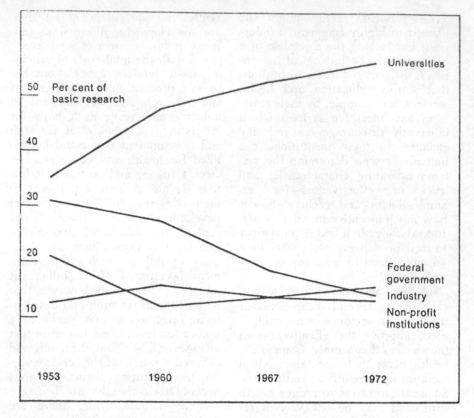

Per cent of
basic research

50

40

30

20

10

Universities

Federal
government
Industry
Non-profit
institutions

1953   1960   1967   1972

Universities have been the traditional sources of basic research in the U.S., but since most of this work has been financed by federal funds it is reasonable to question how much of it has been sig- nificant as a source of new industrial technology. The author suggests a basic research program funded jointly by in- dustry and government. (Data: National Science Foundation)

tive arrangements.

In addition to the services of gov- ernment, health, and education, there are services connected with in- dustrial products; these, too, are costly and growing. The repair, maintenance, and disposal costs of durable consumer goods are high and increasing. Here, too, incen- tives to design more reliable and longer-lasting products are pos- sible—not only to reduce repair costs, the consumption of raw ma- terials, and the pollution generated

in manufacturing, but possibly to create new markets. The federal government could, for instance, pur- chase radically new products and require that they be maintained and disposed of by the producer. This method has the net effect of making the producer actually provide the *service* that the product is to render —reliably, and over an extended and definitive period of time. Consider- able technology already has been developed for the military and space programs that could be used in the design of more durable consumer products. Currently, however, the producer and original buyer are concerned primarily with the initial product performance and have no way to adequately anticipate future repair, maintenance, and public disposal costs. Encouraging the sale of a consumer service rather than a consumer durable might provide a better way to produce goods at reduced total cost to individuals and to society.

Option 5: Support Training and Relocation of Displaced Workers

Workers are displaced and sometimes the economies of whole regions are depressed as a consequence of technological change. While there is no evidence that technology reduces employment in the long run or in the aggregate, change obviously causes local and often severe individual and social dislocations: workers with particular skills are displaced and may not find other employment; regions and cities, like Appalachia or Seattle, become economically depressed. The costs of the technical change are borne by the small number of people affected, while the benefits of the change flow to the society as a whole. Growth industries do have some incentives to retrain and relocate their workers as they expand to serve new markets and reduce their overall costs. However, in slowly growing industries, where the total number of jobs in the industry may be declining, displacements are particularly costly. The mere threat of displacement, as well as displacement itself, slows the innovative process and reduces further the competitiveness and growth of these industries. The textile, shoe, and fisheries industries in the United States are likely examples of this phenomenon.

Publicly supported retraining and relocation programs reduce the inequities induced by technological change and stimulate the overall process of change. There are today a number of such programs, some supported publicly and some privately. However, a more broadly based program paralleling those established in Sweden, West Germany, and—recently—France could be highly effective in the United States. France's imaginative new program, which might be called "educational welfare," requires the worker and his employer to contribute to an educational fund vested with the employee. The accumulated fund, after a certain initial period, may be used by the worker for training or education that improves his skills or makes him more adaptive to a new job. Such a fund, which could also be used to pay relocation costs to another region where jobs are more plentiful, encourages adult education and retraining; and the firm, and society, and the worker all share in the cost of retraining as they share in its benefits.

Even excluding such a scheme, a major review and overhaul of present training and relocation programs in the United States seem desirable. The legislation now under consideration that benefits only particular classes of workers, such as displaced scientists and engineers, does not seem equitable. Since the consequences of changes in public policy or technology affect many workers, an equitable program would be one that treats all workers alike and allocates the costs between worker, employer, and the public at large.

Option 6: Support High-Risk Ventures

A number of studies indicate that a disproportionately large number of major inventions and innovations come from private inventors and small, innovative firms—especially research and development firms. The private inventor and the small firm alike face the extraordinarily difficult task of obtaining early development support for ideas that have not yet proven to be feasible technically or to lead to saleable products. While the individual inventor or new firm must bear or find support for initial costs, the costs of inventive and innovative activity in large and profitable firms can be treated as an expense within the present tax structure; such costs are deductible as future benefit costs, and, in a sense, are partially supported by a reduction in the taxes of the firm. No similar tax benefit flows to the private inventor or the fledgling firm not yet in business; however, there are some loss-carry-forward benefits if the firm "survives," and these could be extended. As suggested earlier, there are arguments to support a tax reduction for incremental research and development, but such a tax reduction would benefit only existing large firms. Yet large firms with large technical and marketing organizations and complex plants often are unwilling to invest in inventions or innovations that are new to their business or that may not be directly applicable to them. The establishment of a negative income tax for the first few years of a new firm's life, however, would give a benefit to the new firm analogous to the tax reduction on research and development for established firms.

A publicly supported organization that provided high-risk, early support to private inventors or small firms in exchange for a share of the equity would also help bring new concepts to the stage where the enterprise system could more readily evaluate the technical and market risks. The judgment for support and the amount of equity should be based on an assessment of opportunity for each case. Over the long run, it is likely that an institution that supported very high early risks could be self-supporting. Because an inventor or entrepreneur usually needs advice concerning the availability of venture capital, business techniques, and marketing, the organization that furnishes the early-risk capital could also arrange to have business schools, which use faculty and others with practical, entrepreneurial experience, provide this expertise.

Option 7: Improve the Transfer of Technology

During the post-war reconstruction of Europe and Japan, the United States had little incentive to seek out new technological developments from abroad or to be concerned about the patent rights and know-how obtained from foreign firms. Indeed, the flow of innovation was presumed to be from the U.S. to overseas users; we provided capital and know-how to Europe and encouraged the manufacture of certain military goods in Japan.

We have limited the flow of technology to the Soviet Union and its satellites, and prevented the flow to the People's Republic of China on ideological, political, and military

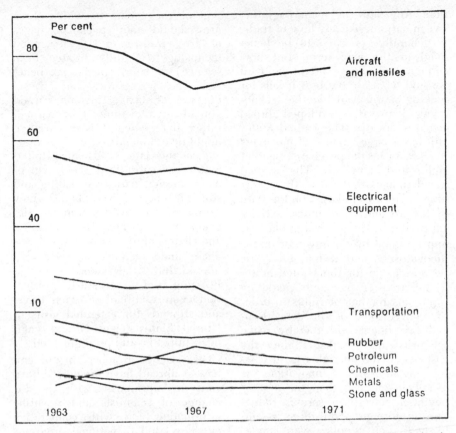

Per cent

80

60

40

10

Aircraft
and missiles

Electrical
equipment

Transportation

Rubber
Petroleum
Chemicals
Metals
Stone and glass

1963   1967   1971

The federal contribution to U.S. research and development varies widely among industries; this chart shows the proportion of all research and development scientists in various industries who were supported by federal funds during the past decade. Should a larger federal commitment be made to U.S. industries where innovation seems to be faltering and foreign competition most sucessful? (Data: National Science Foundation)

grounds. Economics have not been seriously considered. During our two post-war decades of leadership in science and technology, particularly in space and defense, we became convinced of our scientific and technological superiority and even stimulated the Europeans—especially the French—to be concerned with an irreducible "technological gap"— a hue and cry which in recent years seems to have been muted.

Now the situation is quite different. The negative as well as the positive consequences of our past federally supported programs are more evident and more carefully considered. We are beginning to understand the consequences of our failure to recognize the need for public policies and programs to improve productivity and to ameliorate the systemic and deleterious consequences of our system of production and consumption. Any flow of technology between nations, like any flow of trade, is generally conceded to be beneficial in the long term. But how foreign technology should be paid for and what are the best means for making more available the technological knowledge developed abroad remains unclear. The United States still has a large balance of payments in its favor for the purchase of patent rights and know-how. The original developments that led to this favorable balance were based on the state of the art and on technical activity supported by the general public and appropriated by the inventor or the innovator. Would it not, therefore, be reasonable for the United States to tax transfer payments made by foreigners for patent rights in order to recover some of the general social investment and thereby make the prices more realistic? Since the cost of purchasing rights to successful inventions or innovations is usually much less than the cost of beginning afresh, a serious reduction in foreign acquisitions is unlikely if such a policy were implemented. Even if there were a reduction, would not the result be equitable and reduce some of the relative foreign advantage?

United-States-based multinational firms internally transfer know-how and also provide training and experience for foreign workers. Through the establishment of foreign branches, a United States firm may be able to enter markets not otherwise open to it and create a demand for U.S. exports that support the foreign production. Furthermore, these firms are able to obtain foreign technology from their overseas operations for use in the United States. Even so, there is the question of the extent to which these firms should enjoy special privileges of "free" transfer of U.S.-developed technology (as differentiated from science) or enjoy special tax benefits. The possibilities and consequences of taxing income earned abroad at the time it is earned, rather than when it is repatriated, might be examined.

Since the late 1940s the United States has subsidized the travel of many foreign nationals visiting and studying here. We should now capitalize on this program to develop new means for obtaining knowledge of the new applied science and technology developed abroad. Industrial associations could be supported to establish information centers abroad. Foreign travel and stipends for extended visits of United States scientists and engineers abroad could be augmented.

The "science attachés" in our embassies abroad have heretofore been concerned primarily with the exchange of scientists and scientific information and with cooperative efforts related to national programs of defense, space, atomic energy, and health. A similar and substantially greater effort to assist in obtaining technology affecting industrial productivity and environmental effects would now appear to be justified. Though only a small number of people would serve in this role, they might act as catalysts to aid in changing the view of U.S. firms and technicians toward the greater exploitation of techniques developed abroad. Like other countries, the United States must choose the fields in which it will concentrate its technical resources, and buy technology wherever possible. In the future, unlike the past, there will be many

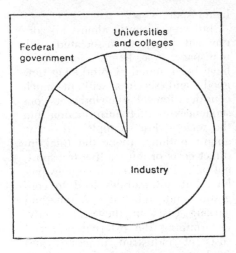

At least 85 per cent of U.S. development work—in contrast to basic and applied research—is being performed in industrial laboratories in 1972. Leading industries in terms of development are aircraft and electrical equipment and communication (both heavily involved in federal work) and machinery, automobile, and chemicals (where the federal commitment is much less). (Data: National Science Foundation)

fields in which one nation or another leads the United States—and we must adjust to this new circumstance.

Through its years of technological leadership, U.S. engineers have developed a "not-invented-here" complex. They don't believe they have anything to learn from foreigners. They would rather reinvent than learn from others. The contrast with attitudes of Japanese engineers is especially striking. The Japanese are prepared to take the best from anywhere and really learn it thoroughly to the point where they can improve on it.

Option 8: Ameliorate the Consequences of Technological Change

We know that a number of major social and environmental problems have arisen as a result of the widespread use of technology and the concurrent changes in the social, physical, and political environments. Serious questions have been raised about the stability of our growing and changing system, about the relationship between private benefits and social costs of technology, and about the ethics and values basic to our society.

The Renaissance, the Reformation, the Scientific Enlightenment, and the Industrial Revolution brought into being institutions and values that encourage individual success, the exploitation of resources, and the destruction of the common environment. Air, water, and land, as well as our aesthetic environment, are polluted as a consequence of uncontrolled individual activities which benefit those who initiate them but which do not lead to the long-term benefit of the society as a whole. The growing recognition of these harmful consequences is reflected in our re-examination of both the institutions and the norms of society—i.e., the literature of dissent.

In principle, the long-range consequences of physical pollution can be estimated by comparing the direct benefits and costs of continuing growth within the existing framework of society with the benefits and costs of technological change. True, the methods of measuring some forms of pollution are unavailable, the consequences unknown, and the possible mechanisms for correction undeveloped; but the basic issue is one of determining the limits of

203

permissible pollution and applying incentives or disincentives to motivate the society to reach those levels.

Three types of activity are required to curb pollution. First, we must determine the harmful side-effects of various contaminants and develop methods of measurement. Second, we must determine permissible pollution levels and the social and economic mechanisms needed to achieve them. Third, we must learn the technology of contaminant reduction, how to substitute new means of production, or how to make products that are less polluting.

There are some who question whether the present economic and political systems can be patched to ameliorate even the physical consequences of technological change and the further economic growth pattern of our society. In the case of aesthetic pollution—such as crowding, devastation of the landscape, ugly billboards, and uglier buildings—the questions are more ethereal and, consequently, less quantifiable. The consequences of psychic distress expressed in the loss of production and increased crime and drug abuse are also only partly determinable. Indeed, these issues of aesthetic and psychic disturbances are more philosophic and religious, related to the deepest values of individuals in the society and of the institutions which they construct.

All of these distressing consequences are interrelated in ways that are little understood. Even so, there now exist activities that might with more public support contribute to the amelioration of the growing contamination of our environment and the destruction of our commonly held resources.

Until very recently almost all government programs associated with non-space, non-defense technology have been aimed at producing new goods and services without much consideration of the indirect consequences of their introduction into the society. For example, we know almost nothing about the total indirect costs of automotive transport; not only do combustion engines pollute, but automobiles lead to congestion and high traffic volume and parking costs in the central city. Determining the external or social costs of industrial processes and products is essential if we are to devise optimum ways of reducing these costs.

Nearly all of the present federal support programs aimed at meeting the energy crisis are devoted to improving the efficiency of or reducing the pollution in the generation and transmission of energy; almost none of the analysis, research and development that might lead to reducing the use of energy is being supported. Because direct energy costs are so low and because energy users are so diverse and diffuse, few incentives exist for technical programs to improve the effectiveness of the use of energy and thereby reduce the waste inherent in the energy production and use system.

As we determine social and external costs and ways of reducing the use of commonly held resources, the industrial system of the United States must be modified to reduce these costs and the destruction of common resources. The changes necessary to the system will sometimes act against the immediate self-interest of those affected and will surely be resisted. Social research and experimentation are necessary to determine those incentives and disincentives which balance long-term interest with the short run.

Changes that ameliorate contamination or reduce the exploitation of natural resources will be introduced by the same processes of invention, innovation, and diffusion as other technical changes. A highly sophisticated base of technical knowledge and appropriately trained people will be required to use this knowledge effectively. Though a great technical effort will be required to quantify the social costs of pollution and set standards that will reasonably balance costs and benefits, the change we need will be effected mostly by "pull" mechanisms. Support will be required for high-risk ventures, for reducing the risks of innovation, and for assuring the widespread diffusion of newly developed techniques.

National programs were initiated in the 19th century to encourage invention, innovation, and diffusion of technology to and for the American farmer. Activities of a similar scope seem necessary now to support the efficient use of products and services by highly fragmented consumers. Just as mining technology has been supported by the federal government for nearly two centuries, we might now develop an equivalent program aimed at recycling waste and reducing the consumption of natural raw material. Methods of monitoring environmental pollution and measuring its effects have begun to receive attention—but not sufficient to change the present pattern of growth. Additional schemes similar to those now being implemented for pollution controls for automobiles are needed to provide incentives for industry to invest in technological

developments in the long-term interest of the society: programs for new institutional arrangements, for increased support of new subjects in universities, and for freeing those able to apply these subjects to the major social problems we face. The present apparatus of government, often designed to encourage and develop the interests of special groups, is ill-suited to carry forward such a major shift in national priorities or technical activities. There is, for example, no overall authoritative agency in government dealing with science and technology that can estimate and ameliorate the short- and long-range consequences of their application to the society. The often-suggested creation of a cabinet-level department for science and technology should be thoroughly considered.

Epilogue

We began with consideration of a changing world that requires new initiatives that might contribute to the continued growth and well-being of the United States. By growth we mean the growing improvement of the quality of life and the establishment of conditions leading to the preservation of that quality in the future. The effort and changes required will be large. We will need more skilled people and greater investments in science and technology. We will need a far greater knowledge of the processes by which invention, innovation, and diffusion occur, particularly in a society in which the free, individual enterprise system has to be altered so as to more automatically preserve the commonweal.

We do not sufficiently understand the mechanisms involved to be certain either of the consequences of our present patterns of growth or the best ways of changing them. We therefore require deeper knowledge of the functioning of our society and of the potential collective effects of individual actions. We have emphasized that technology and science, if we are to survive, require a growing appreciation of new values and new norms and a fuller appreciation of the self in all of us. It is in a time of change and uncertainty that fresh opportunities arise.

Suggested Readings

Cartter, Allan M. "Scientific Manpower Trends for 1970-1985," *Science*, Vol. 172, pp. 132-140, April 9, 1971.

Ehrenkrantz, Ezra. "School Construction Systems Development—The Project and the Schools." Publication of the Educational Facilities Laboratory, Inc., of New York City, 1967.

Freeman, Richard B. "Effects of R & D: Social and Private Rates of Return; Investment Opportunities," in National Science Foundation, *Alternate Federal Policies Affecting the Use of Technology: Supporting Studies*. Supported by N.S.F. Grant GQ-5. Unpublished.

——. *The Market for College-Trained Manpower*. Cambridge, Mass.: Harvard University Press, 1971.

Hollomon, J. Herbert, and Alan E. Harger. "America's Technological Dilemma," *Technology Review*, Vol. 73, No. 9 (July/August, 1971), pp. 31-40.

Kendrick, John W. *Post-War Productivity Trends in the United States*. New York: National Bureau of Economic Research (in press).

Myers, Sumner, and Donald G. Marquis. *Successful Industrial Innovations*. NSF 69-17. Washington, D.C.: National Science Foundation, 1969.

National Academy of Engineering. *Technology and International Trade*. Washington, D.C.: U.S. Government Printing Office, 1971.

National Science Foundation. Division of Science Resources and Policy Studies. *A Review of the Relationship Between Research and Development and Economic Growth Development*. Washington, D.C.: U.S. Government Printing Office, 1971.

Nelson, Richard R., ed. *The Rate and Direction of Inventive Activity: Economic and Social Factors*. Princeton, N.J.: Princeton University Press, 1962.

——, M. J. Peck, and E. D. Kalachek. *Technology, Economic Growth and Public Policy*. Washington, D.C.: Brookings Institution, 1967.

Organization for Economic Co-operation and Development. *The Conditions for Success in Technological Innovation*. Paris: O.E.C.D. Publications, 1971.

Stobaugh, Robert B., *et al.*, Harvard Business School. *U.S. Multi-National Enterprises and the U.S. Economy*. Washington, D.C.: U.S. Department of Congress, January, 1972.

Terman, Frederick E. "Supply of Scientific and Engineering Manpower: Surplus or Shortage?" *Science*. Vol. 173, pp. 399-405, July 30, 1971.

——. "The Process of Innovation: A Review of Some Recent Findings," in G. W. Wilson, ed., *Technological Change and Economic Growth*. Bloomington: Indiana University School of Business, Division of Research, October, 1971.

Wolfle, Dael, and Charles V. Kidd. "The Future Market for Ph.D.'s," *Science*. Vol. 173, pp. 784-793, August 27, 1971.

On
making the future
safe
for mankind

E. J. MISHAN

A CLOSE study of history might
yet uncover periods during which there was, no less than in our own
day, trepidation, obsessive soul-searching, and persistent reappraisal
of contemporary manners and institutions. There may, too, have been
times like our own in which people looked back with cynicism and
nostalgia and looked forward with exhilaration and apprehension. But
I doubt whether the prevailing anxiety has ever spread so wide, or
whether the sense of something awry, of "something rotten in the state
of Denmark" has ever before reached so far down into all strata of
society and agitated it at every level.

The phenomenon need cause no wonder if we bear in mind that
humanity today stands, not at the edge of one crisis, but at a confluence
of three crises—technological, ecological and social—all obviously related and all engrossing our attention over the last decade. For the
development of mass media ensures that, whatever the specific forebodings, no segment of society can be deprived of its due share in the
general alarum.

The pace of technological advance is the primary fact and is itself
responsible for the other two crises: an ecological crisis currently

THE PUBLIC INTEREST, Summer 1971, 33-61.

dramatized by the phrase "the population explosion," but whose chief visible manifestation is the spread of industrial pollutants over land, sea, and air; and, arising largely from the mounting frustrations of urban life, a social crisis, a seemingly chronic restlessness and discontent marked by such familiar symptoms as the growth of drug addiction, of wantonness, obscenity, and incipient violence. Add to these sources of apprehension the emerging vision of an automated, computerized, highly programmed, and centrally controlled society on the one hand, and, on the other, the existing balance of terror between the great powers, each searching relentlessly for new weapons of yet more incredible destructiveness, and it becomes evident that the phrase "living on the brink of annihilation" is today no idle hyperbole. The crisis, or the conjuncture of crises, is all too real, all too fearful, and quite unprecedented.

It is my conviction that the continued pursuit of economic growth by the "advanced" nations is itself almost wholly responsible for the crisis. The general acceptance of such a view by society, I need hardly remark, would have far-reaching implications for the conduct of our economic policies. Once convinced of the close connections that exist between economic growth and the less amiable features of our civilization, we could no longer anticipate an eventual improvement in our condition by the simple expedient of moving with greater or less momentum along this familiar path. We would think differently, we would act differently, and eventually we would live differently.

Historical complacency

Let us ask an apparently naive question: If it is true that society finds itself at the confluence of these crises, why is it that we have been so tardy in recognizing it? Two closely connected reasons suggest themselves. The first concerns the entrenchment in our over-sized societies of existing institutions—political, economic, technological— and the apparently irresistible momentum they set up toward further economic and technological development. The second reason, which I take up immediately, covers those aspects of the ideology of perpetual progress which, on the basis of confused thinking, lend themselves to complacency—to the belief that there is really nothing to be alarmed about after all. This complacency expresses itself in two main forms, the historical and the scientific.

Those exuding the historical form of complacency use ridicule, most of which turns rather monotonously on such epithets as "Doomsday prophets," "Modern Jeremiahs," and "latter-day Cassandras"—forgetting, perhaps, that Cassandra was invariably right. More important, however, is their use of history as a means of dispelling concern. Talk of unprecedented happenings and they will immediately quote you some historical parallel. *Plus ça change plus c'est la même chose* is the refrain, at once cynical and comforting.

For at least two centuries, they will point out, men have distrusted machines. But, I would say, they were not always wrong. The first half of the 19th century was, in Britain, a time of acute distress, suffering, and degradation for the laboring classes—men, women and children. Whether such an epoch, with its evocation of the "dark satanic mills," was a necessary condition for the material advance of later generations is doubtful. If it was necessary, there is still a question as to whether this kind of inter-generational distribution of costs and benefits is politically moral. But what is more pertinent is the fact that the "industrial revolution" which began in the 18th century or earlier, far from abating, is gathering force and, propelled by the boundless ambitions of technocrats, is expanding over the earth on a scale that has begun to fissure the physical environment and to produce complex chains of ecological disruption. The forebodings of the past may after all soon be vindicated.

The complacent historian may observe that for at least four centuries men have looked back wistfully to an earlier age and deplored the growing materialism and irreligion, the unnecessary bustle and change, etc., etc. From this observation one may deduce a number of things, but *not* that the present age is no more materialistic, no more irreligious, or no more rapidly changing than any other period in history. Since the age of Chaucer historians have been delighted to discover—in poems, essays, sermons, plays, diaries, and novels—a recurring nostalgia for the times when nature was more abundant, communities more intimate, and life more wholesome. In particular, they will find a recurring dismay at the disappearance of the green forests and of the irreplaceable beauty of the English countryside. And if today conservationists inherit this mood of concern and deplore the rapid erosion, since World War II, of coastline, of meadow, dale, and woodland—much, that is, of our remaining scenic heritage—one may legitimately infer that the concern of some people at the destruction of the rare and the beautiful is one of the abiding characteristics of humanity. One *cannot* infer, however, that things have not really changed, and have not changed for the worse in this respect. One cannot deny, moreover, that the remaining area of accessible natural beauty is but a tiny fraction of what it was during, say, the 18th century—when the population, incidentally, was less than a fifth of the present population, and the size of the towns was such that wherever a man dwelt or worked he could be in the open country within a few minutes.

Again, the belief that the end of the world was drawing nigh has been widely held at different times in human history. But from this historic fact there is no consolation to be gained. Only since the last war have men succeeded finally in prying open Pandora's box, and among other exciting things which flew out was the secret of instant annihilation of all living things. Time, measured only in short years, will disseminate this sort of knowledge among smaller, poorer, and less stable nations, some of which are ruled by adventurers or fanatics. From this prospect alone one may conclude that the chance of human life surviving for another century is not strong. To annihilation as the result of human irresponsibility, from military mischance or bluff carried too far, must be added the chances of extinction of our species from uncontrollable epidemics caused by the deadlier viruses that have evolved in response to widespread application of new "miracle" drugs, or from some ecological calamity caused by our inadvertent destruction of those forms of animal and insect life that once preyed on the pests that consumed men's harvests. In sum, doomsday fears of yesterday had no rational basis. Those of today have plenty.

Scientific complacency

The second form which complacency takes is one that is most congenial to the forward-looking scientific spirit, one that implies a view of man's destiny best summarized by Arnold Toynbee's thesis of "challenge and response," a view that is itself strengthened by a belief in the infinite adaptability of man. This was Sir Peter Medawar's view in his address to the British Association in 1969, during which he assured his audience that "the deterioration of the environment produced by technology is a technological problem for which technology has found, is finding, and will continue to find solutions." Admitting that there are difficulties technology will have to overcome, Sir Peter ended with an affirmation of faith in the beneficent potential of science and dismissed the faint-hearted and the doubting Thomases as follows: "To deride the hope of progress is the ultimate fatuity, the last word in poverty of spirit and meanness of mind." Splendid language—though it smacks more of hubris than of faith. Faith speaks in a humbler key.

But however we rate this peroration, the question is surely whether we are to be guided by faith at all. Never in history did we need faith less and agnosticism more, an agnosticism that must encompass also

210

the scientific attitude and the implicit judgments of science. Indeed, irreverence must go further if we are at all in earnest. Not even the pursuit of knowledge for its own sake can qualify as a right in the agenda of a sane society. Unless such activity is motivated and constrained by decent and humane ends, it must remain suspect. Contrary, then, to the prevailing ethos, a vaunted thirst for knowledge is no more laudable than a perpetual greed for possessions. On the authority of the Bible we have it that "He that increaseth knowledge increaseth sorrow." Though the revival of learning and the pursuit of knowledge can indeed make life more interesting and pleasant, it can also outreach its usefulness and become an obsessive activity, gathering pace and extension irrespective of its effects on society. More forthrightly, the growth of scientific knowledge over the last four centuries, channeled into a thousand specializations, and translated into technology by market forces or state power, has become subversive of civilized living.

Let us return in a less accommodating mood to this invitation from a renowned scientist to repose our hopes for a better future in the further advances of science, and to believe that technology will itself solve the problems bequeathed to us by its widespread application—problems such as burgeoning populations, atmospheric pollution, traffic-choked cities, oil-fouled beaches, aircraft noise, and so on. The first thought which *should* occur to us is that such problems can also be solved by the use of less technology, not more. By reducing the production and the use of certain kinds of technological hardware, say the automobile and the airplane, we can certainly diminish atmospheric pollution, tourist blight, and traffic congestion. It may be conceded that there is nothing clever about solving the problem in this way. It goes without saying that science and technology would much prefer opportunities for further research with the aim of discovering ways that will enable us to absorb yet more of these technological all-sorts while limiting the extent of their unwanted overspill. There remains, however, the substantive issue: Which is the better way of relieving humanity of any one of the currently unpleasant byproducts of applied science?

Even if it could be ascertained in advance that science does have a contribution to make in solving some of the problems it has inadvertently brought into being, the relevant considerations in any political decision to finance the required research ought to be, first, the time during which humanity, hanging grimly onto its hardware, has to suffer before substantial relief is at hand and, secondly, the degree of risk incurred in any technological solution to specific problems—and,

211

for that matter, the degree of risk incurred in providing us with new technological opportunities—of accidentally releasing on our heads a plague of *new* ecological or other "spillover" effects. In the light of the experience of the last 50 years—in view particularly of the marked tendency of technological innovation to put at the disposal of every person, sane or sick, moral or immoral, powerful means of (inadvertently) annoying or threatening the health or lives of others—the alternative of seeking an improvement in our living conditions by using less technology rather than more has to be taken seriously.

There are also some consequences of applied science that cannot be undone—the holocaust of natural beauty as well as the risks from rising levels of radioactive pollutants lodged in the air, under the earth, and beneath the seas. Science is unlikely to be able to reduce this risk in our lifetime, a risk that grows with the proliferation of nuclear power generators. And what of the irrevocable damage perpetrated in the name of scientific advance? Does technology plan to restore the lives of the 130,000 people killed each year in car accidents? The man who makes his discoveries available to an imperfect society, a society known to be suffused with ignorance, impatience, avarice, and corruption, may not disown the responsibility for the outcome. Above all, we must not overlook the fact that the existing balance of terror is a direct product of applied science. For without the advance of science, the power for destroying all life on earth many times over (and in a variety of increasingly hideous ways), a power possessed already by several countries, would just not have been possible.

Let us not, then, be too easily soothed by the assurances of those whose bright vision of the future—comprehending as it does a cornucopia crammed with research grants—remains undimmed by those follies of the past and present that could not have been perpetrated without technological progress.

On human adaptability

As for the adaptability-of-man thesis, on which the technocrats place so much store, two questions arise: Can man adapt, and should he?

Man as a distinct species has not changed for 100,000 years. Mentally and physically he is the same mammal that ran through primeval forests in search of prey. Until scientists induce mutations that will transform him into a different being, man can adapt only within limits. It is altogether possible that many features of our new, technology-based civilization move strongly against the grain of man's instinctual needs. Such a civilization might then be unstable, inasmuch as it imposes on ordinary people increasingly intolerable strains, giving rise to such familiar symptoms as a break-up of families, a growing incidence of sex perversions, and increasing recourse to drug-taking, destruction, freak cults, and violent forms of protest, self-assertion and defiance. Let us say simply that recent findings in medicine, in zoology, and psychology are not at variance with the hypothesis that the spread of technology is pushing men beyond the elastic limits of their adaptability.

But even if, at some cost and with the aid of some providential new drugs, man could be made to adapt, *should* he be made to adapt? To hold that man ought to adapt himself—in order "to meet the challenge of the future," as our technocrats so quaintly put it—is surely an inversion of ends and means. A moment's reflection on the theme of the good life suggests that we seek first within ourselves to discover what is good and satisfying for man, and then adapt technology to that end; *not* the other way round. The fact that problems will arise in reaching accord about the essential ingredients of the good society does not weaken the force of this dictum. For the mere idea of continually altering man's way of living in order to fit into a world shaped increasingly by the intoxicating visions of technocrats—the mere idea of society as a sort of residuum, as a by-product of perpetual technological innovation—ought surely to be repugnant to us.

Is economic growth a good-thing-in-itself? I now return to my thesis that the chief cause in the West of society's present crisis is to be found in the economic growth that our leaders are so anxious to achieve. Put less provocatively, my contribution to the debate over this topic will consist of examining, in an informal way, the effects on our well-being of the products and processes of economic growth in technically-advanced countries.[1]

To express a doubt about economic growth as a good thing may seem to some a piece of gross impertinence—it certainly would have seemed so a couple of decades ago. After all, the effect of economic growth, it is commonly asserted, is simply that of making available to men alternatives they could not hitherto afford. Provided they are always free to choose what they wish, how can anyone with a liberal conscience allege that they are not better off?

Now it so happens that the economist himself is best equipped to

---

[1] Though I shall not dwell on it here, the growth of population is not excluded from my terms of reference. For economic growth is one of the preconditions of the secular growth in population.

sow the seeds of doubt and distrust about the value of economic growth. And though I speak for myself only, and make no pretense at being strictly detached from the inquiry, I shall initially be making use—proper use, I hope—of some familiar bits of economic thought. Moreover, I shall first separate the environmental problems arising from economic growth, which to some extent *can* be remedied, from the social problems of economic growth that seem to be more intractable.

We can approach the environmental problems by touching upon some elementary but by no means trivial propositions; for instance (1) that the production costs of a business concern may be greater than or (more often) less than the costs borne by the economy as a whole, or (2) that from the fact that people demand a particular good, one cannot justify its production, inasmuch as people's choices are subject to constraints that are institutionally determined. To explain why there are in the United States 100 million vehicles on the roads, we should remind ourselves (a) that they are all priced far below their social costs, (b) the physical environment created in response to the automobile can make it all but impossible to survive without one, and (c) that the public has no living experience of an alternative and viable non-automobile environment.

In this connection, the economist's concept of a "spillover effect" is crucial. If what I wear or use or produce *directly* alters the well-being of other people—and not *indirectly*, through price changes—I can be said to generate spillover effects. Clearly spillover effects abound in society and range from the trivial (Mrs. Smith's envy of Mrs. Jones's new earrings) to the tragic (the death of one's parents through a traffic accident). The current application of this concept, however, to all forms of environmental pollution introduces difficult problems of equity and allocation.

Insofar as the spillover effect in question is fairly simple, and in principle measurable, the economist would tend to favor excise taxes on the polluting products, or else a scheme of incentives to use preventive techniques (e.g., purification plants), rather than outright prohibition or direct controls. But though the economic literature of spillovers, both of the theory and the applied work, is fascinating and indeed central to those problems of universal pollution which, rightly, have begun to agitate society, I will not discuss it here—save to point out that there are instances of the widespread adoption of industrial products that generate effects so elusive and intricate, or so prolific and interconnected, that the idea of research designed to measure them all, or to deal with them adequately by any of the conventional

214

methods, is chimerical. Two examples will illustrate this important thesis: the automobile and television.

The automobile

I once wrote that the invention of the automobile was one of the greatest disasters to have befallen mankind.[2] I have had time since to reflect on this statement and to revise my judgment to the effect that the automobile is *the* greatest disaster to have befallen mankind. For sheer, massive, irresistible destructive power, nothing—except perhaps the airliner—can compete with it. Almost every principle of architectural harmony has been perverted in the vain struggle to keep the mounting volume of motorized traffic moving through our cities, towns, resorts, hamlets, and, of course, through our rapidly expanding suburbs. Clamour, dust, fume, congestion, and visual distraction are the predominant features in all our built-up areas. Even where styles of architectures differ between cities—and they differ less from year to year—these traffic features impinge so blatantly and so persistently on the senses that they submerge any other impressions. Whether we are in Paris, Chicago, Tokyo, Dusseldorf, or Milan, it is the choking din and the endless movement of motorized traffic that dominate the scene.

I need not dwell on our psychological dependence on the automobile. It is the very staple of automobile advertisements to depict it as a thing with sex appeal, to depict it as a status symbol or as a virility symbol. And, over the decades, as the automobile population has grown, along with vast industrial empires that produce and cater to it, the annual sales of new cars has become a separate indicator of the 'prosperity' of the economy. We have, that is, mesmerized ourselves also into the belief that we are economically dependent upon the automobile.

Our physical or environmental dependence upon this vehicle is, however, in fact the direct result of its adoption. Our cities and suburbs have, in consequence, expanded without pause for the last quarter of a century, and have promoted a demand for massive road-building projects that encourages the flow of traffic—which, in turn, further promotes the demand for traffic projects. Because the motorist wants to see everything worthwhile from his motor car, the choicest bits of the countryside tend to be built over. The motorist wishes to "get away from it all" and the highway-builders, in the attempt to provide him

---

[2]*Technology and Growth: The Price We Pay.* (1969)

with the means to do so, succeed ultimately in ensuring that it is virtually impossible to get away at all. And believe me, people do need to get away. The one economic activity showing really impressive postwar growth is the creation of places we all want to get away from.

One could go on, for the extent of the automobile's subversive influence is unlimited. Robbery, crime, violence all today depend heavily on the fast get-away car. Motorists kill off other people at the rate of 130,000 a year (55,000 a year in the United States alone), and permanently maim over a million. Through the emission annually of millions of tons of foul gases the automobile's contribution to sickness and death from cancer and from bronchial and other disorders is just beginning to be understood. What, in contrast is already fully understood—but about which, for commercial reasons, nothing at all is being done—is the connection between air and automobile travel and the greatest holocaust of natural beauty since the beginning of history. The postwar tourist blight has ravaged the once-famed beauty of almost every resort along the coast-line of the Mediterranean, and much of the hinterland besides.

And not only has the physical environment and the economic structure of each Western country been transmogrified to accommodate this infernal machine, but inevitably also our whole style of life—the sort of food we eat, the clothes we wear, the way we court, the forms of entertainment, all bear its stamp. Indeed, our speech, our manners, our health, and our character have been moulded, cramped, distorted in order to maintain the momentum of an industry whose chief visible achievement has been to transform a society of men into a teeming swarm of motorized locusts that have already eaten the heart out of their towns and cities, and now scurry hungrily over the captive earth along bands of concrete spreading in all directions. The better life we overtly aspire to—and the ease, space, leisure, beauty, and intimacy that are conceived as essential features of such a life—can never be realized in the automobile economy.

In view of their far-reaching and inter-related influences on modern society, a proposal to evaluate the full range of the spillover effects of the automobile cannot be seriously contemplated. A large political decision is called for: Either continue to build roads and automobiles until "something gives," or in some degree to de-escalate —that is, to promote a changeover from private to public transport and to direct the resources released from automobile production and maintenance to the rehabilitation of our cities, suburbs, towns and villages.

The case of television

Communications media, in particular television, produce effects on society that are not easy to evaluate. This is so, not only because they are pervasive and intangible, but also because, even if they lent themselves to measurement, the relevant comparison is that between those habitually exposed and those not exposed to these media—effectively, then, a comparison between present and past generations of the same age group in a hypothetically unchanged economic and cultural milieu. Such an experiment cannot in the nature of things be undertaken, and we are therefore thrown back on informed conjecture—in this instance about three sorts of effects: those on language, on personality, and on the family.

1. Because mass media dispose daily of torrents of words and images, the image-creating resources of a Shakespeare could not hope to meet their insatiable demands. The repeated attempts to compel attention on matters large and small issue in near frenzy. Words are misused, abused, over-used, broken up, incongruously combined. And the sheer volume and interminable repetition is itself destructive of the beauty of language. Words of delicate sentiment begin to lose their fragrance. Phrases once rare or solemn, poignant or poetic, to be uncovered only on particular occasions, get dragged about in the dust of sales campaigns, rolled in with crude imperatives, until they became stale, misshapen, and shorn of the joys of evocation. Even obscene utterances, once reserved for special circumstances, have become so common that they have lost their power to shock or amuse us. Along with the general degradation of language goes the degradation of our response to it.

Moreover, mass media being themselves large-scale manufacturers of popular jargon, verbal fashions sweep the country. Half-consciously, people grope their way toward some voguish cliché, at once to avoid the effort of thought and to produce evidence of being *au courant*. But for every piece of jargon adopted, for every "in" word, a score of fine distinctions are discarded. The rich resources of language fall into desuetude. The aim, once associated with a classical education, that of giving precision to one's thoughts, of imparting dignity and beauty to the flow of one's discourse, is perhaps obsolete for a high-pressure technological civilization—in which, apparently, time becomes scarcer in proportion as labor-saving devices become abundant.

It is hard to believe that a consequent frustration of this nature—the growing inability (and the awareness of that inability) to express oneself fluently and persuasively—has no significant effects on people's character and behavior. Is there no connection between this media-induced frustration and the modern accent on the "image," on the action rather than the word, on "doing one's thing" rather than on expressing one's thoughts?

2. Television is commonly spoken of as an "educative force of immense potential." Without troubling ourselves to evaluate this rhetoric, we may accept the fact that today panels of eminent personages and a diversity of specialists use the medium to address themselves to aspects of politics, science, economics, crime, sex, history, literature, housing, health, art, music, ethics, bringing up the children, and the education of parents. Neither need we enumerate the alleged advantages of these programs—how they enable people to perceive all sides of an issue and to acquire tolerance, if not scepticism. We need touch only on one consequence. The sort of tolerance a man acquires from being witness to continual re-examinations of fundamental questions about religion, politics, psychology, manners and morals, is the product of uncertainty rather than of enlightenment. The distinctions between good and bad, truth and falsehood, vice and virtue, sickness and health, are blurred and reblurred by an unending succession of specialists, victims themselves of the current erosion of the moral, aesthetic, and intellectual consensus on which a civilization is raised. Inevitably, then, the confidence of both educated and ordinary people in their own judgment and sense of right begins to ebb. The tolerance that emerges is the result largely of moral paralysis.

3. Finally, and contrary to superficial opinion, television regarded as an institution must be accounted a potent factor contributing to increased isolation. Allow the programs to be ever so "enlightened," the charge still remains. For it acts to displace our dependence on other people for amusement and affection and to transfer this dependence to the meretricious flicker of the television screen. It saps the authority of parents, and interposes itself between members of the family—maintaining the peace only by disrupting the flow of feeling between them.

It is to be noticed, in particular, that the claims made for this as for other technological innovations relate primarily to efficiency—television seen as a universal purveyor of entertainment and instruction. The associated social losses are in terms of less tangible but more fundamental things, in terms, that is, of a total response to life. They include a stunting of our emotional life, a thinness in our human relationships. I confess that I cannot see sociologists, psychologists, or economists, agreeing on methods of computing these sorts of gains and losses, or coming up with an ideal tax to reduce their ownership or use. Though a political decision is conceivable, I cannot believe that, within the ethos of the existing consumer society, prudent regard to any later consequences of unchecked indulgence in technological knick-knacks would carry much political weight.

With respect to spillover effects in general, therefore, we may conclude that the economist's concept does surely have heuristic value; it does contribute to organized thinking about complex social problems; and it does act as a check to the mood of indiscriminate abuse against "the system." Moreover, under restricted conditions, the skilled economist can integrate the concept into practical programs for making worthwhile improvements. There remain, however, a number of flagrant instances in which the resulting spillover effects are too pervasive, intangible, or complex, for the economic calculus to cope with them effectively. Remedial action, if any, in such cases must depend ultimately upon political initiative.

Growthmanship

Before turning to social problems, however, we do well to appraise the validity of the popular contention that economic growth, whatever its defects, is yet necessary if we are to have sufficient resources available to solve these urgent environmental problems.

The appeal to necessity can, and indeed does, take many forms. A government spokesman can assert (fallaciously, I may add) that only a higher rate of growth will enable the country to improve its balance of payments or check its current inflation. He may also assert, and this sounds more commendable, that it will enable us to help the poor and underprivileged at home and abroad. Mention the slums; mention the shortage of hospitals and staff; mention the schools; mention the plight of the orphans and the aged, and the pat answer is "more economic growth." Talk about the congestion in the cities, the spread of suburbia, the growth in diseases of heart and lungs; talk about the pollution of air and water, the ecological breakdown—and what does the growthman reply? Why, more economic growth of course! By way of illustration let me quote a passage from Anthony Crosland's recent Fabian tract. (Other excerpts are printed in Current Reading.)

Even if we stopped all further growth tomorrow we should need to spend huge additional sums on coping with pollution: it will, for example, cost hundreds of millions of pounds to clean our rivers of their present pollution. We have no chance of finding these huge sums from a near static GNP any more than we could find the extra sums we want for health or education or any of our other goods. Only rapid growth will give us any possibility.

There is something almost exhilarating about the uninhibited opportunism of our growthmen. If, dimly and belatedly, they have begun to perceive an environmental problem, they make use of it on the spot to update the relevance of the historic dogma of growth. All facts become grist to the growth mill. For, wherever economic growth appears to have improved living conditions, we surely have evidence of the benefits of economic growth. *Per contra*, wherever it appears to have made living conditions pretty hideous, why again there is a clear case for economic growth in order to remove the hideous features!

Quite apart from the two-headed-penny character of growthmen's arguments, the call today for faster economic growth in order to tackle our environmental ills is fallacious for at least five reasons.

1. Over the last twenty years the prevalent type of industrial growth, in particular the growth of chemical products, plastics, automobiles, and air travel, generates incomparably more pollution than is eliminated by private and public expenditures. What is more, economists anticipate much the same sort of industrial growth over the next ten years or so.

2. As a slight acquaintance with economic concepts makes clear, a successful prevention and reduction of any specific form of pollution uses up less in the value of resources than it confers in benefits. "Real" GNP, that is, becomes larger, not smaller. Any contrary impression is the consequence of too literal an interpretation of official statistics that, at present, attaches positive values only to man-made goods and ignores altogether the losses arising from the man-made 'bads.' I might add, in this connection, that certain "radical" writers who, in the name of social justice, attack the growing concern with environmental quality as purely a middle-class value—a gratuitous insult to the working man—in the belief that it retards economic growth, are guilty of a compound confusion. Not only do appropriate anti-pollution measures add on balance to "real" income (under conventional economic criteria); not only does investment in more attractive and more variegated environments provide the vital choices that can make an invaluable contribution to social welfare; but the *distributional* effects of such environmental improvements are decidedly progressive. For it is the rich alone, at present, who are able

to opt out of any environment that is sinking in the scale of amenity: not the working man, and certainly not the poor, who have no choice. at all.

3. If in recognition of the social dividend to be gained from pollution-reducing expenditures, a government were indeed to commit itself to use a large proportion of the annual *increment* of GNP to combat existing forms of pollution, the argument for pressing on with economic growth might take on a semblance of plausibility. But the bulk of the annual increment of GNP is at present spent on the usual technological hardware and software. In the United States, for instance, the annual growth in GNP ranges between $26 billion to $50 billion.[3] Of this massive increment, what proportion is directed by the government into *additional* expenditure on anti-pollution activities? No one has yet come up with a reliable figure. But I should be surprised if at present it exceeds one-tenth of one per cent of the annual increment. It is, therefore, up to those who persistently invoke this argument to state the proportion of the increment of GNP that will be directed to attack pollution problems. Until then, the public will continue to suspect that future expenditures on "cleaning up the environment" will continue to fall far short of the damage caused by growing GNP.

4. The "need" for more GNP in order to do good in this and other ways is pure fantasy. True, economists have not yet been so bold as to produce from the available statistics that proportion of GNP which in reality only goes to making life more costly, or to estimate those proportions of GNP that could reasonably be classified under such broad categories as "expendables," "luxuries," "regrettables" "near-garbage" and "positively inimical," but the trend toward larger proportions of such items is unmistakable. Granted that the average American was materially comfortable about 1950 (producing then more per capita than is produced in Britain today), we should hardly feel unjustified in imputing a goodly proportion of this per capita increment of 'real' income over the two decades to expenditures on these unprepossessing categories, with much of the remainder being spent in ways that only make life more costly, frantic, and wearing. Consider for instance, the postwar expenditures on the fantastic build-up of urban and suburban areas all over America along with the accompanying fume and din, the longer hours commuting, the increase in tensions, frustration, and conflict and the consequent ad-

---

[3] The federal revenues alone are growing at an annual rate of between $15 billion to $20 billion.

ditional expenditures on tranquilizers, drugs and medicines, on police, prisons and sanatoria, and on research into the growth of violence, delinquency, and nervous diseases.

To talk then of the *need* for more resources before pollution problems can be effectively tackled is manifest nonsense. It is true only in the trivial sense which accepts as unalterable data all existing institutions, mechanisms, and political programs—among other things accepting as a datum the annual expenditure in the United States of a score of billion dollars on the task of endlessly persuading consumers to buy more of a virtually unlimited assortment of goods that presses hard against the consumers' capacity to absorb them, an assortment ranging from plastic gew gaws to private planes, from liquor to extra automobiles, from electric boot brushes to pornographic literature and entertainment.

5. Perhaps most relevant of all, very little increase in public expenditure is called for. What is called for is effective legislation that puts the burden of curbing further pollution squarely on the shoulders of the polluters. The outcome of such legislation would be a re-allocation of resources away from pollutant-creating goods and toward investment in research for pollutant-preventive techniques.[4]

But if economic growth is not necessary, it may yet be desirable.

Growth and well-being

We shall now face the crucial question: What grounds are there for disbelieving that economic growth—once we are well above subsistence levels—can add to our well-being? Although this appears to be a broad quasi-philosophical question, some specific reasons can be mustered first for doubting it, and second for believing that on

---

[4]The slogan that "he who pollutes should pay the cost of his pollution" is not the same thing as effective anti-pollutive legislation, and can indeed act to encourage delaying tactics. It can, for example, be asked what the costs really are, on whom does the damage fall, and which firms or people can be held to be responsible for what part and with what degree of probability. A case is then made for more research into this complex question and we are back in square one. The proposal above can, in principle, be made specific. It sets a time limit, say three years, after which a specified range of pollutants will be prohibited *entirely* (above some specified degree) unless the enterprise has a permit, renewed annually, entitling it to some greater degree of pollution. One condition under which such a permit might be given would be the existence of unanimous consent of all affected parties (for example in the case of noise). Another condition might be that the enterprise is employing an approved preventive technology and/or that it reduces its output to a level determined by reference to standard economic criteria. Any infringement of the law would bring on action by the public prosecutor.

balance continued economic growth in the West will act to *reduce* well-being. Let us look at these doubts from an economic perspective and consider three propositions.

1. *The oft-quoted "widening horizon of choices" refers only to the range of manufactured products and services.* Ignoring wholly the man-made "bads"—ignoring that is, our inability to escape unscathed from the pervasive and damaging by-products of the manufacture and use of many so-called "goods"—which we have touched on, the horizon of choices has nothing along its expanse corresponding to *the conditions of work.* Not every change in these conditions of work has been for the worse; one can easily think of periods of history over which improvements, say, in factory conditions were enjoyed by many groups in the working force. But there is precious little social choice operating at this end of the market. The pattern of production alters continuously in order to meet the changing pattern of the public in its capacity of consumers. In addition, the specific methods by which goods are produced are not chosen by the workers but simply follow the pattern of technological advance.

Thus, from one period to the next, workers at all levels may find their work more rewarding or else more boring and frustrating. But such responses in an "efficiency-oriented" economy have virtually no influence on the resulting pattern of production or on the techniques employed. All we can say in general is that the idea of work as a source of legitimate pride, as a source of gratification, forms no part of the ethos of an industrial civilization and has no influence whatsoever on the direction it takes. Yet who can deny that, like our environment, the sort of work that men do and their attitude toward their work are among the chief components of human welfare?

2. *The consumer who is observed to reject the customary batch of goods still available to him in favor of a new batch can be regarded as better off* only *if his tastes remain unchanged.* There is no necessary connection between new tastes for new products and being better off. To establish this connection, we have to assume that new tastes are inherently superior to old ones.

When economists address themselves to broad policy questions, they tend to overlook this critical provision—enumerated in every good economics textbook—or to make the convenient assumption that tastes do not change very much. For really poor countries, we might let that go. But for really rich ones, the assumption of given tastes is untenable. To conclude as much, we need not belabor the

223

distinction between "natural" and "artificial" tastes, or that between "spontaneous" and "induced" changes in demand. We know that the difference between the sorts of goods our forebears made use of and those we make use of is not simply a quantitative one, nor even a qualitative one in a narrow sense. From the mere fact that we can indeed buy a horse today, but instead we choose to buy a car, we cannot infer that we are today better off. A hundred years of product innovation has changed the world we live in out of all recognition, and has thereby changed the social context in which choices are made. A real choice would be that of alternative *social contexts*—between, for example, a more leisurely pre-industrial world of small towns, wood fires, mansions and cottages, a close-knit society of privileges and obligations, and, on the other hand, a highly competitive post-industrial world of congested highways, unquiet skies, metropolitan overspill, and the daily scramble for status. But in the nature of things such choices cannot be offered to us.

No less pertinent is it that the tastes we acquire are themselves the products of a changing social context and of the institutions it gives rise to. We all recognize that, in the mixed economies of the West, substantial resources are employed, not in order to satisfy the current pattern of wants, but in order to change them. Resources, that is, are used not to satisfy wants (as the earlier textbooks would have it) but expressly to create dissatisfaction with what we have. Again, there is no need to enter the debate on the efficacy of commercial advertising in moulding people's tastes. Speaking only of the broad social repercussions of commercial advertising, one can hardly deny that it does appear to have succeeded wonderfully in one of its aims—that of making people discontent with what they already own. Indeed it is hard to imagine anything that would throw the American economy into greater disarray than a religious conversion that made most Americans perfectly contented with their material lot.

3. *Once subsistence levels are behind us, the satisfaction we derive from a good depends, not only on the amount of it we ourselves buy, but also on our observation of or our beliefs about what others buy* —this being no more than a formal statement of the notion conveyed by the phrase "keeping up with the Jones'," or of what economists sometimes call "consumer independence." The potency of this "Jones effect" in reducing the satisfaction to be had from a steady rise in per capita "real" income, or in productive power, should not be underestimated. It is doubtless very grand owning two cars when most people around you have not even one. A lot of this satisfaction,

however, evaporates when almost everybody in the neighborhood also has two cars or more. Indeed, a distinguished economist by the name of James Duesenberry took the argument to its logical conclusion in 1950 and formulated a "relative income hypothesis" which states simply that what matters to a person in a high consumption economy is not so much his "real" income as his position in the over-all structure of incomes. In its strongest form, this hypothesis would imply that, given the choice, the affluent citizen would choose say a 5 per cent increase in his income alone to the alternative of participating in a 25 per cent all-round increase of incomes. The evidence in favor of the hypothesis in its strongest form, though plausible, is not conclusive. But in a modified form it is hardly to be controverted. The more truth there is in it, however, the more futile as a means of raising welfare is the official policy of promoting economic growth.

Along with the "Jones effect" operating to dissipate the pleasures of rising real purchasing power within a country, there are other factors that cannot be presumed to be beneficial. Without distinguishing them at this juncture we may provisionally assess their potency by reference to common impressions. For instance per capita "real" income in the United States in 1946 was about half that in 1970. If such an index has any welfare content there should be no doubt whatever that, on balance, life in America today is more enjoyable. But there is plenty of doubt about this among Americans themselves. For that matter, to take a more extreme instance, the average per capita real consumption in America today is about five times as high as it was in the Britain of 1950. But, despite the rationing, my recollection is that life was far more comfortable and pleasant in the Britain of 1950 than it is in the America of today—especially in the large cities.

Growth and the positively inimical

Having, with the aid of some familiar economic ideas, given reasons merely for doubting the existence of a positive relation between economic growth and social welfare, at least in wealthy economies, I turn now to the second part of my task: to distinguish some of the features inhering in technological development that appear positively inimical to society's welfare. Systematic study at this level of inquiry is virtually non-existent. Perforce I move into this area of conjecture with less assurance, though with no less conviction.

I confine myself here to three consequences of scientific and technical progress on the shape of society.

225

Consider first the impact on religion and morality, on custom and tradition, of the progress of science and technology. If men ever *want* to believe in a personal God; if there is in men an instinct for worship; if ever they would be God's creatures, they must henceforth be denied. For the myths by which men live cannot survive the relentless scrutiny of science. There is no mystery, no source of exaltation, no beatific vision, through which men may hope to communicate with God that science cannot turn to ashes. There is nothing preternatural or remarkable that it cannot explain in terms of atmospheric effects and chemical processes.[5]

Humanists have declared for a morality founded on an enlightened consensus rather than on Biblical injunction. Their aspirations stem from the belief in the perfectibility of men. But there is no evidence that the human race is drawing closer to moral perfection or that a new enlightened consensus is emerging to fill the present vacuum. As traditional moral codes crumble before the tide of scientific advance, what is there left in a commercial society to moderate the scramble for material status or to curb the frenzy of self-seeking—attitudes that are at once the outcome and the pre-condition of sustained economic growth in the West?

Indeed, not only is morality involved, but every norm that guides society in its choices. For the Western growth-economy requires a consuming public whose tastes are severed from traditional notions of excellence, a public whose acquisitive impulses are unrestrained by any standards of propriety. Once disjoined from tradition, tastes become the slave of fashion, and fashion the creature of profits. And if such an "ideal" consuming public—uprooted, free-floating, volatile, infinitely mouldable—is conveniently coming into being at a time when the greatest threat to the growth economy is the flagging momentum of consumers' expenditures, thanks are due not only to the hard-working ad men but also to the technocrats who trumpet forth the exciting idea that perpetual and accelerating change is the essence of the civilization we are about to enter; a civilization, that is, in which social norms are to have no time to form and in which there-

---

[5]There is little solace to be had from observation that after centuries of mutual antagonism, religious denominations are drawing together. For this is happening at a time when religion plays no vital part in the organization of society. In a vain attempt to appear relevant to the needs of society, the churches are directing their appeal less to the spiritual life of the community and more towards its social and psychological needs. In adapting themselves to modern life in a bid for physical survival, they are divesting themselves of spiritual authority. They are transforming themselves into lay institutions offering society material aid, guidance, and practical advice. They have little choice in the matter. How many Protestant churchmen today believe in God?

fore conceptions of good and evil, of right and wrong, can only be functional and ephemeral.

There is, of course, a disarming frankness about this vision. For we are not being misled for a moment into the belief that, *given the choice,* men would really opt for this somewhat convulsive form of living as being by far the most gratifying to man's instinctual nature as it has evolved over hundreds of thousands of years. We are simply instructed that to think of such a choice is to think the unthinkable.

Of machines and men

Turn next to the perennial question of machines displacing men. I am not one of those who believe that the original Luddities were wholly wrong. Whatever our judgments are on this episode of history, however, the unfolding of events imparts a new twist to the argument. We know that scientists and technologists busy themselves today producing improved translating machines, machines that can play chess or that can learn to play chess or any other game, machines that can write poetry or compose music, machines that can make complex decisions, evolve hypotheses, and produce mathematical proofs. Of course, we should all be very proud! Is not man truly wonderful! But the exclamations of pride do not dispose of the resulting problem. For ignoring the possibility of genetical innovations that will produce for us a race of superbrains, how will the ordinary man and woman respond, knowing that in one accomplishment after another they can be outdone by contraptions of wire and batteries. For almost everything a man can do, there will be a machine that can do it as well or better, and infinitely faster. Such an aim at least provides one of the great steering lights of technological innovation—an aim, apparently, that technology has little difficulty in realizing.

Adapting his mode of living to the technology of industry and to the flow of gadgets on the market, the man in the street, into every year that passes, sees himself more and more a bewildered spectator to what goes on about him. True, his leisure may increase over time and there may be goodies a-plenty in the supermarkets—a robot and a computer in every home, information unlimited, three-dimensional television, round-the-clock synthetic entertainment, trips to the moon and to the bottom of the ocean. But what of his self-respect? For scientists, technocrats, and professional men there will still be opportunities for distinguishing themselves, though the pace of obsolescence of knowledge is sure to place them, too, under increasing stress. The plight of the ordinary mortal, however, is seemingly inescapable. If his muscular and mental exertions, if his manual skills, come to have no value in a world of increasingly sophisticated computers and elaborate control mechanisms, how can he not feel himself to be expendable? How can he hold his head up when it is plain beyond doubt that as a producer he does not rate; that nobody depends on him for anything; that he is but a drone in a world become a buzzing hive of technology?

Of technology and the good life

We now come to the third consequence of scientific and technological progress on the texture of society.

There can be many reasons why a new product commends itself to the buying public. It may promise a novel experience (as does a private plane or a bugging device) or a new form of home entertainment (as does television) or increased leisure (as does a washing machine or an electric knife-sharpener). Once it becomes universally adopted, however, social consequences emerge which cannot be undone by any single person acting on his own. One fairly obvious outcome is worthy only of ironic comment. No people more than the Americans are addicted to labor-saving devices, and no people are more concerned with their weight and general physical condition. Theirs is an irresistible compulsion to buy anything that saves muscular effort. At the same time, no other country is so lavishly endowed with gymnasia and weight-reducing sanatoria. No country sells so much home exercising equipment and slimming contraptions.

Far more serious however, is the consequence of technological innovation on the relationships between people. If we accept the view that (above subsistence standards at least) the chief sources of men's satisfactions reside, not in the goods they buy, but in such enduring things as love, friendship, tranquility, and the perception of beauty, the question arises: Do the innovations produced by the technology of a growing economy act to promote or to thwart these prime sources of satisfaction? I do not think there can be two answers.

Increased mobility is not a force making for increased friendship, least of all when one is for the most part incapsulated in one's automobile. A person can extend the number of his acquaintances indefinitely without really caring for any of them. A week in a mountain hotel, a package tour in the Mediterranean, may have its moments of elation—though the borderline between elation and anxiety

228

is sometimes difficult to define. But there is time enough only to throw postures, to go through the conventional motions of revelry, and hope that something or somebody will turn up. The flurry of emotions, the stylized infatuations, that such opportunities offer to the young, and the would-be young, do not have any affinity with that serenity of spirit with which I am concerned. As the late Somerset Maugham observed, "We never know when we are happy; we only know when we *were* happy." And part of the reason is simply that (*pace* the American Declaration of Independence) the pursuit of happiness is a fruitless enterprise. For the state of happiness is one that cannot be directly realized, least of all by trying. It appears, to quote the late Aldous Huxley, "only as a by-product of good living." And good living, in this context, means neither fleshpots nor sanctity, but something akin to Plato's ideal of harmonious living. It presupposes a dependable institutional and moral framework held together by common beliefs; one that establishes an external order that does no violence to man's internal order; one which permits his instincts to range without hurt to himself or others. In particular, a society congenial to man is one that strengthens his roots in the earth and makes him a part again of that eternal rhythm of nature in which there is time enough for things to grow slowly; in which there is time enough for trust between people to form; in which there is time to learn to care, and time to wonder and to perceive beauty.

If there is any truth in these reflections, it should be apparent that further economic growth predicated on accelerating technical change can only take us farther from the good life. Can one reasonably hope for an easy, open-hearted relationship with one's fellows in a highly competitive and mobile society, where work has become an endless struggle for material rewards and status? The indispensable ingredient of such a relationship is mutual trust, a quality that was nurtured in the pre-industrial small-scale society held together by overt mutual dependence. The resulting intimacy arising from this close personal interdependence, is the first casualty of technological growth. For in the unending search for greater efficiency, technology seeks expressly to emancipate men from direct forms of dependence on one another. Machines come to mediate between them, and they come to depend ultimately for their wants, not on the care of others, but on a row of buttons and switches.

Unavoidably, then, technological progress provides men increasingly with the elegant instruments of their mutual estrangement, and thus constricts further the direct flow of understanding and sympathy between them.

It is time to stand back and take our bearing in order to determine what, if anything, can be done.

If I may caricature economic ideologies, though not excessively, the 19th century was one that eulogized thrift. The good citizen was something of a miser. The second half of the 20th century has reversed those economic ideals. The good citizen is now something of a glutton. For the declared objectives of increasing material prosperity and expanding industrial output can be maintained only by the vigilant cultivation of virtually insatiable appetites. The distributional injustices associated with the system in no way detracts from this conclusion. Whatever the distribution of the national product—be it perfectly egalitarian—the continued growth of the economy would still require insatiable appetites of its citizens.

To be blunt, economic growth in the West is in fact institutionalized greed. We go on as we do from acquired habit, and from institutional momentum—and because we really don't know how to stop. We fear to jettison our growth ideology, along with our hope of salvation by science, because we can see nothing to replace it; because there is no road back. So we drift on making a virtue of necessity, calling out for more speed, and soothing our apprehensions with technological fantasies, and our consciences with repeated promises to do good works as we become richer.

Where then can we turn for guidance? To the desperate expedient of perpetual guerrilla warfare directed against "the Establishment?" Though I have the greatest sympathy with those actively disliking many of the features of this emerging civilization, I cannot agree that it is *repressive* in any familiar libertarian sense. I can agree, however, that we experience today far greater *frustration* than we did. This is so, not only because (as I have indicated above) the demands of a highly technical civilization and the intuitive needs of ordinary men are beginning to pull in opposite directions, but also for simple political reasons. With the unprecedented growth in technology and population over the past 200 years, political power has continuously gravitated toward the center. Yet it is just because of the extending power and compass of modern technology that centrally-determined policies have such far-reaching repercussions on every aspect of a man's private and working life. The tentacles of government reach into every nook and cranny. Unfortunately, as the political constraints on our freedom of action grow, so also do our personal desires to influence events at home and abroad. For it is today the devoted task of an army of ambitious newsmen and commentators to impress us with a sense of urgency and involvement in events both far and near. But even when account is taken of the political effectiveness of organized lobbies, the influence the ordinary citizen can hope to exert on national issues is all but negligible. The vast populations, and the diversity of interests and beliefs in the large countries of the affluent West, go far to ensure that much of the resulting legislation will be a compromise that pleases few and irritates many.

The sense of helplessness is perhaps particularly keen among the young and impetuous. Yet attempts by extreme groups to sabotage "the system" by direct acts of violence are as ill-conceived as they are immoral. If such violence does spread as to pose a threat to society, the response will of necessity be repressive. For security comes before liberty. Indeed, it is the precondition of liberty. And if perchance the violence cannot be contained, the resulting anarchy will pave the way not for a Utopia but for a despotism, a despotism which, by wielding the immense power of modern technology, can be made all the more onerous and totalitarian.

Turning to more traditional sources, has the political philosophy of someone like Edmund Burke anything to offer us? I think not. His reflections are apt enough for a society in which technology has limited impact. Under conditions of *slow* technological change one can indeed argue a presumption in favor of existing political and social institutions and defend a reluctance to introduce any radical alteration without much forethought and debate. But in a society shorn of its myths, bereft of any guiding ethic, a society that is in the throes of a technological upheaval, the experience of the past has little to contribute.

For much the same reason the works of Karl Marx cannot provide us with clues. However his interpretation of history is appraised, as a guide to the future it must be discounted. For in the last resort it is not the capitalist class that is the villain we have to fear. The villain is technology itself irrespective of the economic system. Moreover, the historical determinism of Marx—the belief that choice is but an illusion, that it is futile to attempt to control or even deflect the preordained movement of history—serves to promote either a resigned or a dogmatic spirit, a spirit comparable to that which sustains the momentum of economic growth. If ever we hope to create a society more congruous with man's nature, the mediating spirit must be the reverse of dogmatic or resigned. It must be pragmatic, reflective, and deeply concerned.

Finally, what of the young? Among those of them who have not yet been sucked into the vortex of the new industrial society, a significant number during the last decade have begun to reject the relevance of the Protestant ethic, as popularly understood. Rejected also by many of the young, in particular the middle-class young, are the alleged economic virtues of the "consumer society" as well as a variety of features associated with what is loosely called "the system." Many other young people, though perhaps less articulate, are nonetheless sensitive to the physical ugliness about them and have ambivalent feelings about the approach of an automated society.

But while these youngsters have let it be known, with varying degrees of politeness, that they do not like "the system," will their ebullience or impulse deliver us from bondage and transport us to the promised land?

The more innocent among them—and these tend to be the more vociferous—are impatient of the facts and of the evidence of their own senses. Being impatient they are prone to take a Jekyll-and-Hyde view of the problem, a view that is common both to evangelical and revolutionary movements. They want to slay the wicked Mr. Hyde—the "system" and its minions, the conspiring capitalists, the vainglorious technocrats, and the corrupt politicians or government officials—and let good Dr. Jekyll live forever in pure love and sunshine. But for better or worse, Jekyll and Hyde are indissolubly wed into the single being, man; and it is for just such mortals that society, the good society we aspire to, has to be fashioned.

It follows that if inhibitions are needed to keep Mr. Hyde in check in order that good Dr. Jekyll can breathe more freely, such inhibitions must be firmly rooted in the unquestioned tabus of society. The search for unchecked release of all instincts, a part of the dream world of the dissatisfied young, and the not-too-young, throughout the ages, leads not to "some white tremendous daybreak," but only to disenchantment and despair.

From infantile visions of being borne forever on the crest of an orgasmic wave no values can be salvaged. Today's unheroic protest movements, hippies, yippies, and others "sore given to revel and ungodly glee" (to quote from Byron) offer no viable alternative to the present dispensation. Their social significance derives only from their being one of the symptoms of the crisis of the West. Hippy colonies, for instance, are not new self-sustaining growths, but parasitic ones—barnacles clinging to the underside of the affluent society.

There is little hope for us in the current trend to what is euphemistically called "permissiveness." Where some affect to perceive increased tolerance, I see little more than a disintegration of sensibility, a failure to distinguish between propriety and impropriety, between decency and indecency, between moral and immoral—one of the unhappier consequences of half a century of unprecedented technological change in the West. Not surprisingly in a commercial society, the sexual aspects of this "permissiveness" have become the most prominent. Year after weary year we are being persistently emancipated from those outrageously repressed Victorians—a figment, if there ever was one—and being persistently urged to escape our sorrows by gorging our eyes on erotic images and lascivious display. The transparent result of these gratuitous acts of liberation is to expose the protesting citizen to increasing dosages of sordid sexual pollution. The blazonry of highly salacious entertainment by cinemas and theatres, the city centres in which rows of shops are given over to the sale of pornographic literature or gadgetry, are sights not likely to promote patriotism, civic pride, or an admiration for the character of one's countrymen. To urge that they "meet a need" is, indeed, a pathetic confession, a virtual condemnation of society.

But one can say more. The "libertarians" or "progressives" who come flocking gaily under the banner of Eros are, unwittingly perhaps, espousing a shoddy cause, a creation of neurotic artists, promoted by ruthless commercialism, and patronized by a shuffling army of Peeping Toms whose quite unnecessary journeys take them further into the jaws of fantasy and further away from the fulfilment that comes only from affectionate communication with others. The search for liberation through pathological excesses, like deranged notions of breaking through some imaginary "sex barrier" or of maximizing potential sensation—notice the resort to the language of technical achievement—are indicative only of frustration and failure, aggravated on the one hand, by the physically debilitating effects of modern urban living and, on the other, by the inordinate expectations sown by predatory sex literature and entertainment. But it is the nature of such fantasies that they can never be consummated, and their deliberate encouragement by such enterprises serves only to increase the risk of exposing the new *aficionados* to unendurable frustrations and despair.

More comment on such doctrines, movements, and fashions, which have no real bearing on the problems that face humanity in the latter part of the 20th century isn't necessary. What *will* have bearing?

233

What, if anything, can one offer in the way of practical advice? I can only reply to this legitimate question by saying that there is preliminary work to be done before we can offer (or accept) such practical advice. Above all, if we are to find a way out of the crisis we must first wrench ourselves free from the dominating ideology of growth. We must instead start thinking about the future in an utterly uncompromising and agnostic way. And, whatever conclusions we are led to, they will have to be predicated on three propositions, each one a judgment of fact.

First, that the earth, seen today as a tiny planet warmed by a dwarf star whirling along in a cold, dark and inhospitable universe, is man's only refuge. In consequence, the notion of a unique and finite globe having limited resources of earth, air and water to sustain the complex ecology of life has to supplant in men's consciousness the older idea of an endless frontier of opportunities for systematic plunder.

Secondly, whatever civilization we choose to adopt, its continuance is not compatible with anything like the current rate of destruction of natural resources by the West or with the current rate of growth of human population.

Thirdly, and as a corollary of the first two propositions, we must persuade ourselves anew that, despite the expanding forces of technology and commerce, the future is *not* pre-empted. Though we all know this to be literally true, and though we are ready enough to accept the belief in free will—at least we act in our day-to-day affairs as if we can choose between alternative courses of action open to us—the temptation when visualizing the future to extrapolate trends is strong in any society that habitually thinks of its history in terms of technological progress. Futuristic studies appear for the most part an exercise in extrapolating scientific and technological trends and then in speculating upon their social consequences. Given, say, specific scientific discoveries between the years 1980 and 2000, or the adoption of specific sorts of technology, the question they ask is: What impact will this make on our way of living? In direct contrast, the new way of thinking predicates itself on free will to the extent of reversing this logical sequence. We are to ask, that is, first what sort of a society do we *wish* to establish, after which the consequences for science and technology are to be determined.

The proposition about free will and its social implications is then no platitude. For such a way of thinking about the future runs counter to that which the West has wholeheartedly espoused since

234

the 18th century. Inasmuch as vast material and intellectual interests in science, technology and modern industry, are deeply entrenched it will be something of a miracle if this new agnosticism comes to prevail in the counsels of men.

Whatever the prospects, it is difficult to envisage any decent way of life without a wholesale reversal of the powerful trends—technological, philosophical, economic—that began in the 18th century. The phenomenal expansion of human population, the secular trend toward centralization, the hectic pace of obsolescence, the spread of automobilization and air travel, the growth in mass media, the increasing mobility and uniformity, all such forces will have to go into reverse if such commonly voiced aspirations as variety, order, intimacy, conservation, care, margin, space, ease and openness, are ever to be realized.

The race that is critical to humanity's future is *not* the conventional growth race, not the pitting of the growth indices of one rich country against those of other rich countries until doomsday. Rather it is the race within each Western nation, or within the West as a whole, between, on the one hand, the existing momentum toward yet faster destruction of the earth's depleting resources and on the other, the slow-gathering forces of sanity and understanding.

Education for the Future

HERBERT J. MULLER

A MONG THE MANY NOVEL IDEAS that soon become commonplaces
at a time of fantastically rapid change is the vogue of futurism.
The future has long been a subject of speculation or dreaming,
of course, but never before have its uses been so insistent as they
have become in recent years. Immense sums spent on research and
development represent a systematic way of shaping it. Specialists
banded in commissions, councils and institutes are busy at equally
systematic efforts to anticipate what it may hold in store; they have
elaborated a score of new techniques for the so-called science of
"futurology." And of late dozens of universities have been offering
all manner of "future-oriented" courses, as the fashionable jargon
has it. Alvin Toffler, author of the best seller *Future Shock*, has
been going up and down the land addressing university audiences
on the primary need of "education in the future tense." Henceforth, he writes, "we must search for our objectives and methods
in the future, rather than the past"—the past that has traditionally
provided the subject matter of education.

Now, Toffler is so completely sold on the new vogue that he
exemplifies its most extravagant tendencies. He writes in a breath-
less journalistic style that may make one wish he would pause now
and then to take a deep breath. He is fond of such coinages as "Ad-
hocracy" to define a basic need of properly future-oriented people,
who may then assist in the creation of the brave new "Super-in-
dustrial" world he anticipates. In education he reflects the popular
faith in technology by welcoming the prospect of "a whole battery
of teaching techniques," including the inevitable computerized
programs. With Ad-hocracy must come provision for "life-long
education on a plug-in/plug-out basis." And so on.

THE AMERICAN SCHOLAR, Summer 1972, vol. 41, 377-388.

Yet I propose to take his thesis seriously, on the assumption that the new vogue is by no means another passing fashion. Toffler is uncommonly well informed about the whole subject, having consulted all kinds of specialists. However naïve, his faith in technology is something that educators have to reckon with, especially since many of them, too, are pleased with all the new educational hardware. Even so, he has many sensible things to say in his diagnosis of contemporary life, including sound criticism of our schools. And even in his extravagances—or just because of them—he forces fundamental questions that are too often overlooked in the universities.

One could say this of his basic proposal that we should look to the future rather than the past for our objectives and methods. To my mind it is obviously no question of "rather than," but of "as well as." In reading Toffler I found myself adhering more firmly to some old-fashioned notions about a liberal education, and rising to the defense of the uses of the past, as represented by not only history but literature, the arts, and philosophy. To these uses, especially for the purposes of value judgments, I shall give the last word. At the same time, much that has come down from the past in the universities is of questionable value for contemporary needs, and has been unquestioned simply because it has long been customary, as comfortable to academics as old habits, and as mindless as the ivy on the college walls. I have myself supported the common complaints of students about the irrelevance of too much of their education, and have pointed to the need of education for the future. This need I think should get the first word.

Then we perforce begin with the commonplaces about revolutionary change and radical discontinuities that have antiquated much traditional policy and practice. Certainly our curriculum has not focused enough on the most relevant use of the past, a better understanding of the extraordinary present, or of who we are, where we are, how we got this way, and where we may be going. Considering specifically our radically new kind of society, with its drive to technological "progress" that has generated our major problems, one has only to ask: How many students acquire an adequate understanding of how the drive got started, how it gathered such terrific momentum and accelerated the pace of change, even though most people do not really like living in a revolutionary world? And why does it nevertheless seem irresistible, bound to continue?

So we move into "futurology," which is naturally based on an extrapolation of current trends. Specialists in it have made popular the year 2000—and be it noted, with nothing like the apocalyptic visions that the year 1000 A.D. once inspired in many Christians. For our students, speculation about what life will be like in that year is not at all idle or academic: they will be in their prime then. To be sure, prediction is necessarily uncertain, for reasons they had better understand clearly. But immediately we have to face the confidence with which Herman Kahn and Anthony Wiener can list a hundred important technological innovations that they say are almost certain to appear by the year 2000. I, for one, am quite willing to take their word for it, inasmuch as scientists and technicians are already busy on the job of realizing these possibilities. Given the common assumption that education should prepare young people for life, one recognizes that it should now make them aware of major developments that are strong probabilities, or even virtual certainties.

This task is all the more important because of the uncertainties about the future, which involve the social and cultural consequences of all the scientific discovery and technological innovation looming up. In view of the compulsive drive to acquire systematically ever more knowledge, and with it more power, the question is: What will man do, or ought he to do, with his new powers? Some of them have obviously beneficent possibilities, which appear in the radiant visions of the future offered by the more naïve scientists and apostles of technology. Others are obviously frightening, for instance, the development of more thermonuclear and chemical weapons. Many have alarming possibilities as means of manipulating people, controlling behavior, even transforming personality. But almost all these powers, including potentially beneficent ones, raise complex social and political problems. It is the problems that most concern serious forecasters, the prospect of still more problems for our policy-makers—who right now, let us add, are not up to dealing effectively with the many grave ones already confronting us.

Hence there is a wide field for courses on the future. Time and space do not permit a review of the many possibilities, but let us take a look at the already popular subject of the environmental or biological crisis, aggravated by the population explosion. Ecologists tell us that if current trends continue we are headed for disaster, conceivably a world that will become uninhabitable. Together with the possibility of a thermonuclear war, these trends make human survival a problem for the first time in history. Meanwhile we are faced with such immediate threats as the increasing pollution and the urban crisis, the steady deterioration of the central cities in the sprawling metropolitan areas. And behind all these problems is the accepted national goal of indefinite economic growth, even though Americans already consume many times their proportionate share of the world's dwindling natural resources. While one may doubt that such growth can continue indefinitely, neither business, government, nor the public is prepared to accept a program of national austerity, which scientists are insisting is necessary. Almost all economists have assumed that steady growth signifies a healthy economy; only of late have some begun to question it. I would welcome courses in economics for the future.

All this brings up a basic problem for educators that troubles me much more than it apparently does Toffler and most other futurists. In a course on modern technology I set students to exploring independently a wide variety of problems, and those who did the most thorough job of research uniformly reported pessimistic conclusions about our prospects. What troubled me was the difficult effort to maintain a nice balance. On the one hand, young people need to be clearly aware of what they are up against, the very real dangers, the good reasons for alarm about the future. On the other hand, they need to retain a hopeful spirit if there is to be any hope for the future. The most that I could say to my students was that the growing alarm was the best reason for hope. They were accordingly heartened by the defeat of the S.S.T. program, the first dramatic success in halting the drive to what most Americans regard as progress. Only they then pointed out that

President Nixon at once assured the public that despite this temporary setback America would retain its technological leadership of the world, and ensure further progress.

In more familiar terms, young people have more need of being flexible, adaptable and resourceful than any generation before them. Education should accordingly help to make them so, immediately through a fuller awareness of their fast-changing world, and then by developing their powers of choice. Some of the new courses on the future, I gather, attempt to do this by games presenting them with possible alternatives. For, although the drive of science and technology will force adaptation to much change, like it or not, they still have plenty of vital options—a point that Toffler rightly emphasizes, in view of the stereotypes about our standardized mass society. More freely than any previous generation they can set their own goals and life-styles, as many have been doing in unconventional ways. As citizens they can hope to have some say about their future by influencing policy-makers. And on both counts I am led to my particular concern—the role of the so-called humanities in education for the future.

So far the specialists in efforts at forecasting have been mostly assorted social or behavioral scientists. Literary men, philosophers, and even historians have contributed relatively little to this whole enterprise, presumably because they are not equipped with the necessary specialized knowledge and techniques. Nevertheless, I believe they have much to offer. We might recall that more than a generation ago, before the vogue of futurology, Aldous Huxley offered an imaginative forecast in *Brave New World*, in which, without benefit of fancy techniques or methodology, he was remarkably prophetic about some developments.

As for formal education, humanities departments could offer relevant courses just because of their traditional preoccupation with the past. An obvious example is a study of utopias, which have been a distinctive product of Western civilization from the Renaissance on, and as a tradition, strengthened by the rise of the idea of progress, have had a growing influence on its history, most conspicuously through Karl Marx's vision of a classless society. Such studies might illuminate the revulsion against the utopian tradition, as in *Brave New World, 1984*, and much science fiction. Then they might make clear that utopianism is nevertheless by no means so dead as it is reputed to be outside the communist world. It survives not only in the apostles of technology but also in such humanistic thinkers as Paul Goodman and Lewis Mumford. Altogether, visions of ideal possibilities—however improbable—are especially pertinent in view of our fabulous technological means.

But my main concern is a more fundamental one. Toffler writes approvingly of "the people of the future," the small minority—even in the advanced industrial societies—who are well adjusted to the increasingly rapid pace of life, in a "short-order" or "instant" culture. They live faster, they regard transience and novelty as normal, they welcome change—in short, they are "the advance agents of man," who are already "living the way of life of the future." Education, it follows, should concentrate on molding many more such people, with specific preparation for changes to be expected. "Even now," Toffler writes, "we should be training cadres of young people for life in submarine communities."

Here I shuddered, and the more so because some of the young people conceivably may have to live in such communities, or in the underground cities envisaged by other futurists. I would concentrate instead on efforts to give the young a better idea of what civilized life has meant and can mean at its best, so that they might help to shape a future in which people would not be condemned to so unnatural a life. The "people of the future" who welcome change are not disposed to think hard enough about the fundamental questions: Is all the change really necessary? Is it *desirable?* If so, by what standards? Such questions raise the abiding issues of permanence and change, of the natural life or the good life for man, of basic human values. While Toffler recognizes the necessity of value judgments in developing powers of choice, I doubt that even whole batteries of the latest techniques can do the job well enough. My thesis is that the most important contribution the humanities could make to education for the future is a basic study of the problems of human values and value judgments—judgments that students are not trained to make in the social or behavioral sciences, whose approved methodologies do not lend themselves to such purposes, but rather support the common illusion that true science is "value-free"; and judgments that call for more knowledge of the past than these up-to-date scientists usually have.

So let us consider more commonplaces. Transience is plainly the order of the day, above all in America, where it has invaded all spheres of daily life. Americans have grown accustomed to impermanence in their surroundings because of their mobility and the constant demolition and new construction; few live out their lives in the houses in which they were born. Their economy features throw-away goods, planned obsolescence, the latest models, the latest fashions; a major purpose of the immense advertising industry is to make them dissatisfied with old possessions. In their cities the seemingly so solid new skyscrapers that keep replacing old buildings are not built for keeps, as men used to build. Lewis Mumford, although a harsh critic of our technological society, has himself celebrated "the death of the monument," in which all societies expressed their aspirations to the everlasting: to this symbol of "death and fixity" he opposed the value of a capacity for vital renewal in civic life. Behind Mumford lay the philosophies of Becoming, from Hegel on through John Dewey, that have largely supplanted the traditional philosophies of Being, based on assumptions of fixities, immutable essences, eternal verities. For Dewey the main end of education was growth—a good word for change. And I should add that a survey of the long ages of man may give one a deeper appreciation of some important respects in which the human race has grown in its pursuit of truth, beauty and goodness.

Yet man could not have grown aware of change, of course, except for a background of unchanging realities. Permanence remains a basic condition of his life in the natural world, including the uniform fate of mortality. As Mumford knew, a capacity for renewal itself implies biological and social constants. These have made possible the enduring values that man has wrested from his endless travails, which have likewise been a constant in his history. Today the extraordinary pace of change may obscure the underly-

242

ing continuities and uniformities, the elementary truth that change is not the only constant in our time. While I happen to cherish old possessions, prefer the old cities of Europe, and admire many of the great monuments of the past, I also write as a conscientious relativist who came to conclude that, short of eternal verities or life everlasting, there are permanent values, absolute goods, which must be considered in any aspiration to the good life—and especially today, above all by "the people of the future."

These goods bring up immediately the issue of the "natural" life for man. The study of history makes plain that we cannot talk easily of such a life for a creature who moved from the cave to the life of the soil, later from the village to life in cities, and who all along developed a remarkable diversity of cultures that alike seemed natural to the people brought up in them. His record shows that man is an extraordinarily adaptable creature. So in adapting himself to an industrial society radically different from all previous ones, he learned among many other things to live by the mechanical clock instead of the natural rhythms of night and day and the seasons. I assume that most people could adapt themselves to life in submarine communities. Conceivably, too, man may in time be made still more adaptable by not only systematic conditioning, such as B. F. Skinner aspires to impose on the whole society, but such imminent possibilities as new wonder drugs, genetic surgery and engineering, and biochemical controls of the nervous system. Another eminent psychologist rejoices in the prospect that by biochemical means "we will achieve the ability to change man's emotions, desires, and thoughts."

Yet "we" would mean in practice the various specialists in social or human engineering. If I am myself too set in my ways, or too content to go on with my own thoughts and feelings, laymen in general cannot simply trust to the wisdom of these technicians. In any case, no educator can stake his all on such future possibilities. Meanwhile we have to deal with young people who have the same basic biological and psychological needs as their ancestors have had through the ages. We must consider too the plain reasons why modern life may in some respects be legitimately called unnatural. Certainly life in a noisy, congested, polluted environment is not natural for man, or good for him, even if people get used to it; quiet, sunshine, fresh air, open space, and greenery are among the elementary natural goods. When René Dubos, speaking as a biologist, calls for a "new social ethic" based on the idea of living in harmony with nature instead of forever exploiting and "conquering" it, we might be reminded that this is a very old idea, common not only in folk cultures but in ancient Greece, Confucian China, Buddhist Japan, and much Western literature. In a historical view it is the technological drive to exploit and conquer, at any cost to the natural environment, that is strictly unnatural, as now its menace to human survival also suggests. So too is the American rage for endless acquisition and consumption to satisfy "needs" created by admen, often at the plain expense of mental health. The more adventurous young people are themselves seeking more "natural" ways of living.

There remain the higher needs of the life of the mind, developed during the history of civilization, which are the special concern of the humanities. Embracing aesthetic, ethical, intellectual, broadly spiritual values, they lead us to the issues of the good life. Needless to say, we shall never agree on this. It would be most unfortunate if democratic educators ever did approach agreement, inasmuch as their task is to assist their students in deciding for themselves what the good life for them is. Then the young people might settle for the life of the superindustrial future as pictured by Toffler. But first I would like to widen their range of choices in possible futures, for both themselves and their society, and to develop their powers of judgment, by making them think hard about the perennial questions, how to live and what to live for—the right questions even though, or again just because, they cannot be given conclusive answers by scientific or any other methods.

Any such effort would call for some introduction to the wealth of possible answers suggested by the diverse cultures of the past, in both East and West, embodied in the great works of art and thought that the human race has hung onto. I doubt that most thoughtful, independent students would welcome Toffler's vision of a superindustrial technology that will manufacture experience instead of mere commodities, since so many are bent on having their own experience, doing their own thing; but for this reason, too, they need to be acquainted with the kind of experience provided by Shakespeare, Rembrandt and Beethoven, and by such rebels against the modern world as Dostoevski, Nietzsche and D. H. Lawrence. In their common quest of significance, ways of ordering life and giving it meaning, they need as well more time sense than is provided by an instant culture that prizes immediacy. With all this they might get too some tragic sense of life, a deeper sense of the abiding realities of the human condition—a spirit lacking in Toffler and in most other futurists, who may, therefore, often seem superficial.

Again space does not permit a survey of the many ways in which the liberal arts may promote a full development of capacities for growth, self-realization and self-determination. (I am assuming that most young people are still not ready to let B. F. Skinner determine the kind of self everybody should have.) Here I would stress in particular the importance of aesthetic education, commonly neglected in both contemporary practice and futuristic theory. In a society whose business is business, it is more necessary than ever before to develop sensitivity of perception, refine powers of choice, and enrich ideas of desirable futures. It could make more meaningful and effective the standard complaints about the quality of American life. Now that President Nixon has added this cliché to his stock, it is fair to remark that he is only typical of our society in that on his record he has a pretty vague or vulgar notion of the standards of excellence the phrase implies, including aesthetic standards, and has displayed little awareness of the deterioration in quality during the phenomenal economic growth of the nation. For young people the aesthetic judgment is more pertinent because it accordingly enters as well into judgment of the life-styles they are experimenting with.

Now, all such old-fashioned talk may seem still more academic and ineffectual because of the facts of professional life in the so-called humanities. I have myself dwelt wearily on the too many thousands of Ph.D. dissertations and learned articles in the professional journals, the great bulk of them written by specialists only for other specialists, with little relevance to basic issues of value judgments, still less to education for the future. For different reasons contemporary art may make talk about permanent values seem as irrelevant. Here transience is the order of the day in the form of the fashions that keep sweeping over the arts, often with the announcement that old styles or art forms are dead. In this welter of novelty there is indeed much imaginative work of merit, perhaps of promise for the future, but most conspicuous is the element of mere fashion, the latest "ism," the rage for novelty. Mostly short-lived, the fashions support Toffler's belief that the values of the future will be more ephemeral than the values of the past.

I suppose he may be right, especially because all the signs are that the pace of change will continue to accelerate. As he writes briskly that "permanence is dead," so the historian J. H. Plumb writes soberly that "the past is dead." But if so, I should still say so much the worse for the future. Meanwhile, at any rate, permanence is not actually dead, nor is the past. We still have plenty of earnest students who are not so wedded to transience and novelty as Toffler's people of the future. The actual problem for educators aware of the challenges posed by the pace of change is much more difficult than fighting a rear-guard battle in a steady retreat. It is to create a more selective, usable past, by instilling more reverence for its great works of art and thought that can still speak meaningfully to us, but also by recognizing its too common tyranny, in both the world of thought and the political world; and by instilling an appreciation of both the enduring values it created and the values of change and growth, recognizing the inveterate unreasoned tendency of most people to resist change, but now also the too common tendency to accept it as unthinkingly. In short, a perpetual balancing act, in which educators can never be sure that their timing and emphasis are right.

They face a related problem of balance in the judgment of science and technology, in particular because of the animus of many literary intellectuals against them. It is obviously futile to treat them as simply a curse, inasmuch as we now could not possibly live without them. Quite apart from all the material benefits flowing from them, they too have inspired great imaginative works, while contributing the indispensable scientific spirit to the values developed by our civilization. Today humanists have no more effective allies than the many life scientists who are deeply concerned over the problems of not only human survival but survival in a decent environment, fit for fully human beings to live in. At the same time, humanists have to contend against the popular faith that science can give us all the answers. Although they may find allies too among behavioral scientists who know better, they have to keep pointing out that too many others have an excessive faith in the latest techniques, or in a methodology that, far from answering the fundamental questions, tends to obscure them or rule them out. Toffler, an ardent pupil of the behavioral sciences, is typical when he calls for a "science of futurism," the establishment of more "scientific futurist institutes," scientific "measures" of the quality of life, social experiments with "the most rigorous, scientific analysis of the results," and so forth. He is too sophisticated to believe that all this could give us positive answers, but the magic word remains "scientific"; so it is necessary to emphasize that he is using it in a not at all rigorous sense.

Finally, however, I would emphasize chiefly that any serious concern for the future demands first of all thoroughgoing criticism of the present state of America, to which both scientists and humanists have been contributing. The root of our troubles is the American way of life, with business as usual, politics as usual, consumerism more than usual. I think that education for life in that year 2000 might do no better than turn out more Ralph Naders.

BIBLIOGRAPHY

HISTORICAL PERSPECTIVE

Hoover, Herbert Clark and Lou Henry Hoover. De Re Metallica.
Dover Publications, Inc., 1950.

Davidson, Thomas. "The Ideal Training of an American Boy,"
The Forum, 17:571-581, July, 1894.

Vaughan, Victor C. "The Bubonic Plague," Popular Science
Monthly, 51:62-29, May, 1897.

Baxter, William Jr. "Forecasting the Progress of Invention,"
Popular Science Monthly, 51:307-314, May, 1897.

Austen, Peter T. "The Utilization of Waste," Forum, 32:74-84,
September, 1901.

Gore, James Howard. "The Metric System and International Commerce,"
Forum, 31:739-744, August, 1901.

Graffenried, C. de. "Condition of Wage-Earning Women," Forum,
15:68-82, March, 1893.

Underwood, Sara A. "A Forgotten Industrial Experiment," New
England Magazine: An Illustrated Monthly, 18:537-542,
July, 1898.

CONTEMPORARY PERSPECTIVE: THE DISORDERED TECHNOLOGY

Hughson, Roy V. "Is Technology the Cause of the World's Problems?" Chemical Engineering, 77:90+, June 29, 1970.

Wurster, Charles F. "As Long as it Doesn't Kill Anybody,"
Innovation, 17:21-28, January, 1971.

Borlaug, Norman E. "In Defense of DDT and Other Pesticides,"
The UNESCO Courier, pp. 4-12, February, 1972.

Neslo, D. "Heresy in the Hinterland," Journal of Industrial
Arts Education, 26:37-39, September-October, 1966.

Craig, Paul P. "Lead, The Inexcusable Pollutant," Saturday
Review, pp. 68-70+, October 2, 1971.

Tollett, Kenneth S. "What Price Ecology?" The Center Magazine,
3:20-21, July/August, 1970.

Hardin, Garrett. "The Tragedy of the Commons," Science, 162: 1243-48, December 13, 1968.

Dubos, René. Man's Participatory Evolution--Is Man Over Adapting to His Environment," Current, pp. 34-39, April, 1971.

CONTEMPORARY PERSPECTIVE: THE ORDERED TECHNOLOGY

Fuller, R. Buckminster. "Commitment to Humanity," The Humanist, 30:28-33, May/June, 1970.

Guilbert, Madeleine. "Women and Work: The Effects of Technological Change," Impact of Science on Society, 20:85-91, January-March, 1970.

Lerner, Max. "The Culture of Machine Living," The UNESCO Courier, pp. 23-27, May, 1971.

Ward, Barbara E. "Women and Technology in Developing Countires," Impact of Science on Society, 20:90-101, January-March, 1970.

Helgren, Fred J. "Schools Are Going Metric," The Arithmetic Teacher, 20:265-67, April, 1973.

Graham, W. Fred. "Technology, Technique, and the Jesus Movement," The Christian Century, 90:507-10, May 2, 1973.

MacGregor, Ian K. "Technology and Society," Vital Speeches, 37:525-29, June 15, 1971.

Adams, Ruth C. "How to Love The Land and Live With Your Love," Organic Gardening and Farming, 19:68-72, May, 1972.

FUTURE PERSPECTIVE

Holloman, J. Herbert. "Technology in the United States: The Options Before Us," Technology Review, 74:32-42, July/August, 1972.

Mishan, E. J. "On Making the Future Safe for Mankind," The Public Interest, pp. 33-61, Summer, 1971.

Muller, Herbert J. "Education for the Future," The American Scholar, 41:377-88, Summer, 1972.